WITHDRAWN

BUILDING
THE
PUBLIC
CITY

URBAN AFFAIRS ANNUAL REVIEW

A semiannual series of reference volumes discussing programs, policies, and current developments in all areas of concern to urban specialists.

SERIES EDITORS

The **Urban Affairs Annual Review** presents original theoretical, normative, and empirical work on urban issues and problems in volumes published on an annual or bi-annual basis. The objective is to encourage critical thinking and effective practice by bringing together interdisciplinary perspectives on a common urban theme. Research that links theoretical, empirical, and policy approaches and that draws on comparative analyses offers the most promise for bridging disciplinary boundaries and contributing to these broad objectives. With the help of an international advisory board, the editors will invite and review proposals for **Urban Affairs Annual Review** volumes that incorporate these objectives. The aim is to ensure that the **Urban Affairs Annual Review** remains in the forefront of urban research and practice by providing thoughtful, timely analyses of cross-cutting issues for an audience of scholars, students, and practitioners working on urban concerns throughout the world.

INTERNATIONAL EDITORIAL ADVISORY BOARD

RECENT VOLUMES

BUILDING THE PUBLIC CITY

◆

The Politics, Governance, and Finance of Public Infrastructure

edited by

DAVID C. PERRY

URBAN
AFFAIRS
ANNUAL
REVIEW
43

SAGE Publications
International Educational and Professional Publisher
Thousand Oaks   London   New Delhi

For information address:

 SAGE Publications, Inc.
2455 Teller Road
Thousand Oaks, California 91320

SAGE Publications Ltd.
6 Bonhill Street
London EC2A 4PU
United Kingdom

SAGE Publications India Pvt. Ltd.
M-32 Market
Greater Kailash I
New Delhi 110 048 India

Printed in the United States of America

ISSN 0083-4688

ISBN 0-8039-4432-2 (cl)

ISBN 0-8039-4433-0 (pb)

95 96 97 98 99 10 9 8 7 6 5 4 3 2 1

Sage Project Editor: Susan McElroy

Contents

Acknowledgments

Some of the essays in this book started out several years ago as contributions to the Albert A. Levin Lecture Series in Urban Studies and Public Affairs. The growing national and local interest in the perceived state of our urban infrastructure combined to make the series, titled "Building the Public City," a very popular set of lectures at the Levin College of Urban Affairs at Cleveland State University. The series and the subsequent contributions to this volume were stimulated by this contemporary interest in public works and, at the same time, were grounded in the fact that the issues of public infrastructure, so central to *present* political discourse, have been with us throughout the history of the country. The issues of economic development, health and safety, and debt (or finance) were as important in the first days of the republic as they are now. This point is made as well today by the National Committee on Public Works Improvement as it was by Adam Smith, more than 200 years ago, when he suggested in *The Wealth of Nations* that public works form the underpinnings of the free market and the public right to health and safety. As we look toward the future, the age-old issues of economy, health, and debt are now translated into issues of economic development, environment, and debt.

These new forms of key developmental issues challenge the ongoing constellation of political, governmental, and fiscal institutions and practices of public works delivery. It is these institutional features of the politics, governance, and finance of public infrastructure that comprise the major topics of this volume.

To the extent that this book succeeds in adding to our understanding of urban policy, the success is due in no small part to the contributions and support of many people. Key among these wonderful contributors are Dennis Judd and Susan Clarke, the editors of the Urban Affairs Annual Review series. Their patient support and constructive criticisms of our work have been instrumental to the final completion of this volume. At Sage, I have come to depend a great deal on the advice and support of Carrie Mullen, a fine and able editor. Also at Sage, I would like to thank Myrna Reagons, Tricia Bennett, and Susan McElroy. At

Cleveland State University, where I was honored to be the Alfred A. Levin Chair of Urban Studies and Public Service and where the whole "Building the Public City" project was begun, I am grateful for the support of the Maxine Goodman Levin family, especially Dr. Maxine G. Levin and Morton Levin. My colleagues at CSU are among the finest urban scholars I know of. I especially want to thank Richard Bingham, David Garrison, Ned Hill, Dennis Keating, Norman Krumholz, Helen Liggett, and David Sweet for their various contributions to this project. At the State University of New York at Buffalo, my home institution, I am grateful for the financial support of the U.B. Foundation and the early direction of Harold Cohen, who helped in the initial stages of the development of the Robert Moses Research Project. The project and parts of this book benefit, as well, from grants from the Triborough Bridge and Tunnel Authority and the National Endowment for the Arts. I appreciate the suggestions of my colleagues at U.B., especially Alfred Price and Ernest Sternberg. Along the way I have benefited from the advice and support of many others, including, in no particular order, Mitchell Moss, Robert Cohen, Alfred Watkins, Dennis Crow, Seymour Sacks, George Palumbo, Alan Campbell, Susan Fainstein, David Szanton, Harold Blake, George Schoepher, Frank Mauro, Richard Nathan, Arnold Vollmer, M. Christine Boyer, and Nancy Connery. Neither the book nor the research would reach completion without the stalwart help of a grand group of students, including Jon Lines, Ellen Parker, Kathleen Mooney, John Shannon, Gilbert Chin, Richard Guarino, and Geoff Poremba. If, however, there are flaws to be found in this book, as there certainly will be, the blame rests with me and certainly not with all these wonderful people who have added so much to my understanding of public infrastructure and urban policy.

Finally I want to acknowledge my family: my sons Clayton and Evan, and especially Judith Kossy, for providing the intellectual and emotional support it takes to help make projects like this a part of the life of the family and not a problem for it.

1 Building the Public City: An Introduction

DAVID C. PERRY

Good roads and canals will shorten the distances, facilitate commercial and personal intercourse, and unite, by a still more intimate community of interests, the most remote quarters of the United States. No other single operation, with the power of Government, can more effectively tend to strengthen and perpetuate that Union which secures external independence, domestic peace, and internal liberty.

Albert Gallatin, U.S. Secretary of the Treasury, 1808

Infrastructure . . . stands for the connective tissue that knits people, places, social institutions and the natural environment into coherent urban relationships. . . . It is shorthand for the structural underpinnings of the public realm.

Herbert Muschamp, *The New York Times*, 1994
© 1994 by The New York Times Company.
Reprinted by permission.

The politics of public infrastructure, or "building the public city," are among the most broad-ranging of any in American political economy. From Albert Gallatin's description of internal improvements early in the nineteenth century, to Herbert Muschamp's claims for infrastructure near the end of the twentieth, to President Clinton's recent promise to inaugurate a multibillion-dollar program to "rebuild America," it is hard to find a more impassioned or purposive policy discourse than that concerning public works.

Yet, for all the heady discussion, public infrastructure is often taken for granted. When roads, bridges, waste disposal sites, and water systems work best, they are noticed least or not at all. And when attention is finally focused on "infrastructure" (usually in response to system failure or severe and continued congestion) real damage has already been done" (National Council on Public Works Improvement [NCPWI], 1988, p. 33). This reactive pattern has certainly been the case for most of the past two decades, with one national study after another describing a litany of decayed and deteriorating conditions plaguing the nation's public facilities (Choate & Walter, 1983; NCPWI, 1988; U.S. Congress, Office of Technology Assessment [OTA], 1990, 1991; Weiman, 1993).

It is both the seminal role that public works play in social formation and their ubiquitous influence in our everyday lives that make an understanding of public infrastructure policy important. Public works are at the very heart of the physical, technological, and fiscal foundations of the American city; or, as urban designer William Morris puts it: "Infrastructure is the safety net of the social system." To be blunt, when the public infrastructure of a city fails, the entire city as well as our individual daily existence can be dramatically changed.

This was never more clearly the case than in Chicago when, early on April 13, 1992, the people of the city woke up to the news that 250 million gallons of the Chicago River had crashed through a relatively small break in a long-abandoned underground tunnel, flooding the buildings of the downtown retail and financial core, causing the evacuation of more than 200,000 people, shutting down a substantial share of the transportation system, and causing upwards of $1 billion in property damage and business and tax losses. All this because city public works crews had failed to carry out "a $10,000 patch job" lamented the *Chicago Tribune*. "Try to control it and a river will break your heart," the newspaper went on. "Reverse its flow, change its course, abuse it with sewage and industrial spillage, and it may sit docilely for decades: old man river, shuffling along. That does not mean it has been tamed, however. It is just biding its time" (Johnson, 1992, p. 1, © copyrighted Chicago Tribune Company, all rights reserved, used with permission). What the paper was really referring to here was not the capriciousness of nature, but the folly and forgetfulness of humans—those who would build a city on an infrastructure that alters the course of nature in the service of urban change, and then would neglect to continue maintaining the public realm they had created. The real problem was not the inconstancy of the Chicago River, but the inconstancy of those who built the public city—the canals, tunnels, buildings, water systems, and electrical power grids that are comprised in the public realm in which their daily lives are carried out.

In fact, the real news here is not simply that the Chicago disaster is one of such rare magnitude, but rather that the poor maintenance, fiscal neglect, and increased urban dependence on decaying and deteriorating facilities represented in the Chicago River flood are *not* so exceptional. For example, it is estimated that about 40% of the 600,000 bridges in the United States are rated as deficient, and on average, 120 of them actually collapse every year; "some," argues Clark Weiman (1993, p. 43), "simply because they never received a new coat of paint and were more vulnerable to the elements." Similarly, a plethora of national studies points to the fact that almost 60% of the nation's highways need repair work, and almost three quarters of American auto owners consider highway congestion to be a major problem. Still other studies suggest that two thirds of the wastewater treatment facilities in the country are failing to meet water quality standards, and more and more urban ground water is showing detectable levels of volatile organic chemicals. The environment is further threatened by landfill sites unable to accommodate the production of more than one-half billion tons per year of solid waste. Both the economic development and the health and safety of urban populations are threatened by thousands of hazardous waste sites.

In part we owe such conditions of obsolescence, deterioration, and neglect to a highly fragmented and shortsighted practice of public infrastructure policy. In spite of a politics in which the primacy of public infrastructure has been recognized since the beginning of the republic, there has never been anything even close to a coherent public infrastructure policy in this country. On the contrary the history of public infrastructure is one of fitful and conflictive federal initiatives, employed more often during periods of macro-level economic crises and technological change than in response to the consistent and recurring felt needs of states and cities.

The lion's share of public infrastructure activity has been carried out at the local level—providing an almost inexhaustible array of fiscal, governmental, technological, and public-private arrangements in order to build, manage, and maintain the full range of public works. Although this book does deal with the "national state" of public infrastructure, most of the work here is concerned with "building the public city": with historical, institutional, and case analyses of the places where public infrastructure policy has traditionally mattered the most—at the state and local government levels.

Even at the local level the actual process of building public works was and is a problematic one. The rapid rate of nineteenth-century urbanization in the United States created conditions that overwhelmed many attempts to deal with such problems (Anderson, 1988, p. 137; Konvitz, 1985, p. 101). The increase in both the number and the population of cities in the nineteenth century created unprecedented levels of disease, congestion, and fires—generating urban disasters of a magnitude similar to the one just described in 1992 Chicago. Historian Letty Anderson has found significant lags in the time it took cities to mount concerted attacks on such conditions. First there was a lag between the appearance of the problem (i.e., disease or fires) and the perception of its true significance. A second lag occurred between the time it took to mobilize people to respond to the problem and the discovery of a perceived solution (e.g., the installation of citywide water systems). Third, Anderson suggests there was always a delay while cities tried to come up with plans to "minimize the uncertainty of the results" (Anderson, 1988, pp. 137-157). Anderson found that these three time lags were reduced as more cities found out that others had applied similarly acceptable solutions to similar problems. Josef Konvitz (1985, pp. 96-106) suggests this pattern of urban intransigence occurred almost every time cities attempted to employ new technologies to meet new issues of internal improvement. He found that

> ideas and techniques in city building did not evolve as rapidly as the economic, demographic and social aspects of urbanization. City builders, anxious to exploit the opportunities urban growth provided, mastered the design and construction of specialized, utilitarian building types (many of an unprecedented scale or functional purpose) by applying already proven techniques to labor, land and capital. (Konvitz, 1985, p. 101)

Put another way, even after it became clear that new infrastructure systems would help with conditions of rapid urbanization and stimulate economic growth, cities would take years to determine the "right approach" and even more to decide on how to pay for the system (Konvitz, 1985; Warner, 1972). For example, although the relationship of water systems and water treatment to the control of disease was fully established by the late 1880s (Anderson, 1988; Tarr, 1988), water treatment, sewers, and sewage treatment did not become everyday facts of life until

TABLE 1.1 Total U.S. Population, Urban Population, Water Treatment, Sewers, and Sewage Treatment, 1880-1940

Year	Total Population	Urban Population	Water Treatment Population	Sewer Population	Sewer Treatment Population
1880	50,155,783	14,129,735	30,000	9,500,000	5,000
1890	62,947,714	22,106,265	310,000	16,000,000	100,000
1900	75,994,575	30,159,921	1,860,000	24,500,000	1,000,000
1910	91,972,266	41,998,932	13,264,140	34,700,000	4,455,117
1920	105,710,620	54,157,973	N.A.	47,500,000	9,500,000
1930	122,775,046	68,954,823	46,059,000[a]	60,000,000	18,000,000
1940	131,669,275	74,423,702	74,308,000	70,506,000	40,618,000

SOURCE: Tarr and Dupuy (1988). © 1988 by Temple University. Reprinted by permission of Temple University Press.
a. 1932 data

well into the twentieth century (see Table 1.1). "Ironically," observes Konvitz, "city building techniques that emphasized economic criteria produced environments ill-prepared to adjust to many of the changes accompanying urban development" (1985, p. 101).

In the absence of any clear federal role in the late nineteenth century, this pragmatic and temporized approach to public works delivery was all the more understandable, given the scale of the projects and the costs of production. The introduction of large-scale networked sewerage systems, for example, demanded large-scale financial investments and geographically broad networks of pipes and pumps, which spread across municipal boundaries. Their installation could be politically perilous in that, although these technologies could remove sewage from one municipality, they might have deleterious "downstream" effects for a neighboring community. More broadly,

> the various characteristics of sewerage technology, such as its capital intensiveness and its planning requirements, as well as the fact that its effective operation bore little relationship to municipal boundaries, required a number of institutional adaptations and innovations. These involved planning innovations for large scale public works, developing methods to raise capital funds. (Tarr, 1988, pp. 178-179)

By the 1940s the technological networks promised by nineteenth-century reformers were finally in place: sewers, water systems, tracks, and electrical and telephone wires crisscrossed the city, providing services at a level that was inconceivable in the 1890s (Tarr & Dupuy, 1988).

Along with the technology and the physical networks, however, other features of building the public city were also in place. The scale of public works in the twentieth century far exceeded any dreams in the nineteenth century, and with scale came issues of government and finance. Building the public city in the twentieth century introduced substantial new issues of governmental relations. The magnitude of the demands for new highways and utility systems, the reemergence of the federal government, and the creation of local public authorities and special purpose governments served as three of the cornerstones of service delivery. The final cornerstone of policy formation was finance: Not only did the new technologies of public works portend substantial changes in the environment, they also caused substantial increases in public debt.

■ Public Infrastructure

Few policy terms have received such a level of new and widespread meaning as the term *infrastructure*. Earlier in this century, infrastructure was essentially a military term used to describe the permanent fixtures of military installations—the base camps and airstrips or fixed ports that, when taken together, formed the systems of defensive and offensive warfare. The historian Bruce Seeley (1993, p. 20) suggests that it was economists who first extended the meaning of the term by including, for example, those features of a society that W. W. Rostow once labeled "social overhead capital" (1960). By this Rostow and others meant the capital invested in roads, utility systems, communications, and education, health, and other governmental facilities as the foundation for economic development. However it was not until the 1980s that infrastructure replaced *public works* as the term most commonly used to describe the physical artifacts and what Joel Tarr and Gabriel Dupuy (1988) have called the "technological sinews" of contemporary social formations. Today the notion of infrastructure has come to include "almost every support system in modern industrial society, public or private. Infrastructure is said to include not only roads and sewers, but national transportation grids, communication systems, media, housing, education . . . computer networks and fiber-optic 'information superhighways' " (Seeley, 1993, p. 20). Or as Muschamp (1994) concludes, infrastructure is "shorthand for the structural underpinnings of the public realm."

Josef Konvitz combines the definitions of Seeley and Muschamp when he concludes approvingly that "unlike public works, which it subsumes, the term infrastructure is at once a description of physical assets and of their economic, social and political role" (Konvitz, 1985, p. 131). And of these three roles, the role most often stressed is the economic one. Most recent descriptions of infrastructure, whether they be ones describing the uses of public projects (OTA, 1990) or ones assessing conditions of physical decay and fiscal distress (Choate & Walter, 1983; NCPWI, 1988), include some measure of its sufficiency to meet "future economic growth and development" (NCPWI, 1988, p. 1). In this sense what Thomas Jefferson and George Washington called "internal improvements" or, later on, what engineers, reformers, and Franklin D. Roosevelt called "public works" are today described as "infrastructure": the technologies and their structural/spatial applications that serve as the support systems of market society. The term *public* is added to infrastructure as a way of referring to the mixture of public- and private-sector activities that go into the definition, design, finance, operation, and maintenance of these support systems.

What separates the notion of public infrastructure from the earlier terms internal improvements and public works is its inclusiveness and its emphasis on systems of support. Earlier discussions concentrated on the improvements—the highway itself, or the canal itself, as part of the support system. And the same can be said of public works. In 1984 Triborough Bridge and Tunnel Authority Executive Secretary Harold Blake said that

> 1984 is the first time I have ever heard the term public infrastructure applied to a city in peacetime. I am sure Robert Moses would have found it quite absurd to apply such a term to what we all just called 'public works'—the bridges, roads and parks we built in New York. (Blake, 1984)

What Blake was referring to was the incremental political practice at the local level that, for much of this century, placed less emphasis on describing public works as systems and more on the bridges or highway networks as projects. But ultimately, the term *public infrastructure*, meaning the "technological systems, with each road, bridge and drainpipe closely linked to an intricate . . . delicate . . . network of supporting elements" (Seeley, 1993, p. 20), is a far more complete description than public works or the nineteenth-century term *internal improvements*. Hence, though the authors in this book will use the terms *public works* and *public infrastructure* interchangeably, the essays mean to convey the inclusiveness (technological, systemic, political, economic, and social) of the term *infrastructure* to these studies of the policies that help build the facilities of the public realm.

Ironically, though the primacy of the relationship of these fragile technological systems to growth and development has a long-standing currency in America, there is a similarly lengthy history of American reluctance to pay for public works (Perry, this volume). For something considered so key to the practices of American privatism (Warner, 1972), it is, at the least, puzzling to discover how problematic and inconsistent the finance of the nation's internal improvements—or public works or public infrastructure—has been (see Anderson, 1988; Foster, 1981; McKay, 1976; Schultz, 1989; Tarr & Dupuy, 1988). For features of the public realm so fragile and potentially capable of breaking down, it is surprising how little time, money, and effort are employed in rehabilitation, maintenance, and upkeep.[1] This contrary combination, of the accepted importance of public infrastructure and the lack of stable finance and maintenance policies to support it, forms much of the dynamic basis of the politics of public infrastructure.

■ Building the Public City

The politics of public infrastructure are at once about the impacts of technological change, the goals of economic and social development, the changing demands of governmental relations, and the recurring crises of public debt. In this sense building public infrastructure represents some of the most profound and conflicted goals and political issues of American policy formation. This multifaceted politics is the topic of each of the contributions to this book. Together they comprise what we call *Building the Public City*.

Technological Change

Infrastructure is deeply embedded in historically significant technological changes—changes that have had important impacts on the overall direction and development of the republic and its cities. The importance of technology to social change rests not in the technology itself but with the way the new technology is transformed into a process or systemic support of social formation (Castells, 1989, Forester, 1987). By the middle of the nineteenth century, many Americans had come to believe technological and physical innovations could solve the health and safety problems of the urban environment. Historian Stanley Schultz described the work of sanitarians, civil engineers, and landscape architects as the dominate nexus of

> the city-making landscape in America. They articulated a new concept of "public works," thereby giving expression to the changing social vocabulary of the century's enraging urban culture. That conception urged that the public itself, through its tax dollars and its political support for public works projects, take charge of its own destiny in building a safer, saner and more sanitary urban environment. (Schultz, 1989, p. 153)

Engineers became the paragons of the public works reform culture—embodying all that was efficient, technologically superior, and devoid of political corruption. If technology was at the center of infrastructural change and reform, the engineer, an expert in technology and the process of design and implementation, was the professional practitioner of political reform and innovation. By the end of the nineteenth century, according to James Olmstead, the municipal engineer was

> responsible for holding the successive political officials to a consistent, progressive policy in all the branches of work under his change. To him, even more than to the successive mayors, falls the duty of serving as the intelligence and brains of the municipal government in all physical matters. (Seeley, 1993, p. 29)

In 1905 a Baltimore engineer used the sewers of Paris as evidence of the importance of the engineer to the quality of urban life:

> Paris, completely sewered, with a low death rate . . . [is] the center of all that is best in art, literature, science and architecture and is both clean and beautiful. In the evolution of this ideal attainment, its sewers took at least a leading part, for we have only to look at conditions existing prior to their construction to see that such a realization would have been impossible before their existence. (Schultz, 1989, p. 153)

The blend of technology and politics as key ingredients in building the public city is the topic of Jameson Doig's chapter "Politics and the Engineering Mind: O. H. Ammann and the Hidden Story of the George Washington Bridge." In this chapter Doig describes, in detail, the way in which the great bridge builder Othmar Ammann combined his design for the gigantic bridge across the Hudson that would ultimately link New York City and northern New Jersey with years of political activity in order to forge a coalition in both states to support the project. In the process Doig introduces us to the internationally acclaimed railroad engineer Gustav Lindenthal, whose practices and designs epitomize the early twentieth-century practices of public works, where the engineer orchestrated policy in a "rational working environment," with strong private sector support, freedom of design, and the public acclaim that came from "doing things right." In order to build bridges or other public works, Lindenthal believed that building public works was not a political process, rather that

> insulation and professional integrity were crucial, and the engineer could be relied upon to work out the best way to meet the other goals and constraints within which all great engineering projects must be designed and carried out—concerns with economy and with the esthetic quality of the project itself, and an understanding of how the project would fit into the broader patterns of economic and social relationships within the region. (Doig, this volume, p. 35)

Lindenthal represented the engineers' acceptance of what historian Richard Wiebe has called the almost "childlike faith" Americans placed in them during the first half of the twentieth century. Ammann represents a substantially different practice—one where

> the substantive arguments and the political strength of opponents [to a public works project] deserved the same steely-eyed analysis that a good engineer devoted to understanding the stresses on bridge cables and the stability of the ground under proposed bridge towers. If influential opponents thought the bridge too large, or badly located, Ammann could draw on his engineering experience and perhaps find ways to modify the plan, rather than let it go down to defeat. (Doig, this volume, p. 35)

As far as financing problems were concerned, Ammann was more than willing to blend public and private monies in practices that hearkened back to the early days of the nineteenth century when, as Carter Goodrich (1960) describes it, public works were "mixed public/private enterprises." In this chapter, Doig describes the emergence of Othmar Ammann as more than simply an engineer and designer. For Doig, Ammann is also a "political entrepreneur." The notion of "political entrepreneurship"—the combination of technology, marketeering, and public interest—goes to the heart of the practice of building the public city.

Doig suggests that the biographic significance of his chapter rests with the fact that this would be the first and last time that Ammann would directly take on the role of political organizer. He would go on to design some of the greatest bridges in the world, including the Golden Gate and many of the bridges serving Manhattan, but he would leave the political organizing to others. Ammann immersed himself in the institutional power of the Port of New York Authority and the Triborough Bridge and Tunnel Authority, letting political operatives such as Austin Tobin and Robert Moses take the entrepreneurial lead. In sum, this chapter points to the necessary relationship between technological innovativeness and political activity as a dual foundation upon which to build the public city.

The role of engineers has diminished in the last third of the twentieth century in spite of a series of federal policies that focused national attention on infrastructure development after the Second World War. First, the 1954 National Highway Act initiated the construction of 45,000 miles of interstate highways. The 1965 Highway Trust fund was inaugurated to provide a revenue stream to pay for the highways. And in 1968 the creation of the Department of Transportation gave Cabinet-level status to transportation issues. The environment received attention with passage of the Solid Waste Disposal Act of 1965 and the creation of the Environmental Protection Agency in 1969. Water quality was singled out with the passage of the Water Pollution Control Act of 1972. Even with these efforts, however, the conditions of congestion, air and water pollution, and crises in waste disposal seem to have continued almost unabated. Debt-ridden mass transit systems and poorly maintained highway and bridges are the legacy of the technological tunnel vision of engineers and the largesse of federal policies that, for most of the fifties and the sixties, highlighted what was purported to be the "golden age of public works" (Foster, 1981, p. 179).

By the early 1970s the combination of urban decline and suburban sprawl in many American cities was as much exacerbated by the solutions of engineers and their public works projects as it was abated. Pollution, debt, and overcrowded highways were enough to temper the public's long-term love affair with the automobile and, along with the first signs of fundamental structural crisis in the industrial urban economy, were enough to shift the policy direction of public works at all levels of government as well. Beginning in the 1970s the public discourse on public infrastructure became deeply embedded in issues of development—especially economic development.

Economic and Social Development

This change in the politics of building the public city is addressed directly by another engineer, Robert Mier, in his chapter "Economic Development and Infrastructure: Planning in the Context of Progressive Politics." Like Ammann, Mier was a practicing "city-builder." For more than 7 years, between 1983 and 1989, he served as the Director of Development for the City of Chicago, during the tenures of Mayors Harold Washington and Eugene Sawyer, and the chapter he contributes here is a description of his experiences in using public infrastructure policy as a central feature of urban economic development. Unlike Ammann, Mier's task was not to engineer the construction of any particular bridge or stadium, but to engineer the overall rebuilding of the urban economy. To do this Mier tied public infrastructure policy directly to urban economic development in ways that set him apart from Ammann and put him at odds with many other public officials. Mier tied notions of economic development to what he terms Mayor Washington's "social justice agenda," representing a "departure from previous public economic development practice and set[ting] a context for strong business interest contention" (Mier, this volume, p. 74).

The bottom line of this agenda was jobs, as reflected in a three-part strategy calling for (a) the expansion of the urban economy to reach "full employment"; (b) new public-private partnerships that included neighborhood and working people in the project development decisions, which heretofore had been controlled by traditional alliances of city hall, business, and labor leadership; and (c) a reversal of the "big project" approach to urban infrastructure development, with a more decentralized neighborhood-based program of public works development.

Central to Mier's approach to public works policy is the notion that the building of public infrastructure is a political process, which, if it is to be successful, must fulfill a variety of objectives comprised in a complex economic and social agenda. Mier uses three different types of public infrastructure projects (a proposal for a World's Fair, a neighborhood-based infrastructure program, and a reconsideration of a large-scale public works initiative) to make this point. Each of these cases offers clear evidence of how contentious the public works environment can be, especially when the political process places issues of social equity on the same plane with issues of economic productivity. Mier's purpose is to show how building public works can serve the twin missions of social justice and industrial development. However, he suggests, although such an approach to public infrastructure and economic development need not be antagonistic, the odds are that it will be.

Both the Doig biography of Ammann and Mier's self-professed "stories" offer what Mier sees as the "real stuff" of the practice of city building. Mier suggests elsewhere that "stories are the bread and butter of development decision-making. Seldom do developers and politicians relate their desires, successes, techniques, and the like, through studies and analyses. . . . Rather, 'war stories,' provide daily staple. Thus it is stories that connect development ideas to development practice" (Mier et al., 1993, p. xi). It is for these reasons as well that Doig's use of biography and Mier's particular brand of storytelling enrich the methodology as well as the substance of this book's treatment of public infrastructure policy.

Claire Felbinger, in her chapter, "Conditions of Confusion and Conflict: Rethinking the Infrastructure-Economic Development Linkage," also addresses the relationship of public infrastructure to economic development. Where Mier describes the complexity of the political environment that links economic development and public infrastructure, Felbinger explores a range of what she calls the "roles" offered to the public, physical systems she describes as public infrastructure. In the process Felbinger discusses what she perceives to be a fundamental shift in the function of urban public works. She argues that the historic nineteenth-century function of public works, as responses to the spread of cholera and typhoid brought on by inadequate water supplies and uncontrolled sewage, quickly grew into important networks linking systems of water filtration, sewage treatment, parks beautification, and street alignments, thereby transforming public infrastructure into the primary source of health, safety, and human convenience. Felbinger goes on to argue that in the twentieth century, the primacy of public works as physical and "human systems" of health and safety has been diminished, replaced by an increasingly economistic and developmental set of roles for public infrastructure.

Felbinger attributes this change in the politics of public infrastructure to the lack of a clear federal public works policy. This lack of clarity at the national level contributes to (a) the decline of the federal role in cities in general and in public infrastructure in particular, and (b) the increase in state and local participation in public works. The result of this, argues Felbinger, is intergovernmental relations of public works that can best be described as "a form of spatial and fiscal triage, where the projects most often built are those that can attract some form of funding and can be built in areas exhibiting the fastest economic growth" (p. 116). These relations reinforce federal fiscal policies, which most readily supply funds for capital projects that can demonstrate a real return in increased jobs (construction) and fixed capital plant. The systems that embed public infrastructure first and foremost in the economic development policy of a region leave conditions of health, safety, and environment in a decidedly secondary position, suggests Felbinger. Such practices, she argues, place the city or region at risk.

In the remainder of her chapter Felbinger sets out to redress this imbalance, pointing out that issues of health and safety are not unevenly distributed across the productive landscape. Felbinger carefully suggests that the health and safety needs of urban and regional populations are not perfectly matched up with economic development issues. Rather, the health-related issues of public works delivery are spread out far more broadly, often reaching their most acute state in areas where the population is relatively light and economic activity is far from dynamic. "Put another way, issues of health and safety, far from being strategically distributed in productive areas of economic development, are spread across the entire urban and rural region and can be translated into broad issues of the environment as well" (p. 117).

Governance and Debt

Issues attendant to the political-economic rationale for producing public works—whether this rationale is situated in a logic of economic determinism or one of health, safety, and convenience—cannot be separated from the pragmatics and politics of governance and finance. Jameson Doig, for example, describes Othmar Ammann's attempt to build the Brooklyn Bridge as a study in "political entrepreneurship," where questions of public ownership and finance were far more crucial than issues of technology and design. In the last three chapters of this book James Leigland, Heywood Sanders, and David Perry direct their attention to the topics of public ownership and public debt in chapters on the governance, politics, and finance of public infrastructure.

James Leigland, in the chapter "Public Infrastructure and Special Purpose Governments: Who Pays and How?", introduces his discussion of governance with a brief review of literature (deHaven-Smith, 1985; NCWPI, 1988; Peterson & Miller, 1982; Porter, Lin, & Peiser, 1987) that "calls on state and local governments to create their own governmental corporations to finance and manage public works services" (p. 139). Leigland suggests that regardless of whether we agree with this literature, special governments (public authorities, boards, corporations, and other types of special districts) are the most rapidly growing category of new governments in the United States. Special purpose governments offer a plethora of new revenue strategies, including earmarked taxes, user fees, bond agreements, and trust funds, which "are said to offer ways of circumventing unreasonable restrictions on public borrowing, increasing the likelihood that construction management and service delivery are carried out more cost-effectively, and shifting the burden of public works financing away from all taxpayers to those directly served" (p. 139). Leigland critically assesses the role of these special purpose governments, providing an overview of the types of governments and financial strategies they provide, along with an assessment of what problems they resolve and what further issues they generate.

By the end of his chapter, Leigland leaves little doubt that special purpose governments play "a crucial role in financing and managing public works provision" (p. 165). Yet he is equally clear that there are serious questions that must be raised concerning the future use of these governments. Leigland concludes that continued dependence on the special purpose government offers increased and essentially unchecked debt formation, furthers the structural conditions of functional overlap and fiscal fragmentation of public service delivery at the local level, and militates against more centralized public infrastructure initiatives from the state level.

In the chapter "Public Works and Public Dollars: Federal Infrastructure Aid and Local Investment Policy," political scientist Heywood Sanders joins Leigland in presenting strong empirical evidence of the complex and substantially differentiated approaches that are taken to public infrastructure policy at the state and local levels. Although

Leigland and Sanders start from somewhat different readings of the data on the national condition of public infrastructure, with Leigland taking a less critical position than Sanders over how to read the national studies and their conclusions of physical crisis and decay, they both provide persuasive evidence that the study of building the public city must begin, if not end, with state and local governments. Though they both offer analytic evidence for expanded and changing roles for federal public infrastructure policies, they both suggest that the most fruitful place to look for institutional, programmatic, and fiscal solutions to current public works issues is at the local level.

Beyond these general levels of agreement, however, Sanders offers a far different treatment of public infrastructure policy from Leigland's. He provides what could be described as a revisionist view of the "national crisis" approach to public infrastructure. In the process he too addresses the question of "*how* we invest in infrastructure and *why* streets and sewers in many parts of the country appear to be failing." He suggests that rather than approach these issues from what he views as an artificially produced set of national patterns, we should focus on the "process of political choice at the state and local levels." He suggests that only after we understand how infrastructure decisions are made— how political choice is manifested—can we begin to suggest what a realistic federal infrastructure policy should look like. In discussing the issue of "political choice," Sanders describes a political process composed of a "complex federal-state-local" partnership that will ultimately benefit from a renewed, informed understanding of how decisions are made and finances are managed. In so doing, Sanders contends that arguments "about infrastructure need and financing requirements [that] stress enormous structural and condition problems demanding a vast fiscal commitment [are] in reality far more modest and manageable." "The rhetoric of collapse and underspending," he argues, "obscures the reality that problems are concentrated in a very few states or cities" (p. 196). He finds further that figures on capital spending are more closely matched than earlier studies suggest. As a result, he sees the principal problems concentrated in major cities, especially older cities. And he sees the federal solution more clearly tied to national urban policy than to specific national public works policy. He concludes on a far more optimistic note than Leigland, arguing that "infrastructure needs are by no means insurmountable. They do require that we make choices about how much we raise taxes and revenues, what our spending objectives are, and that we recognize the complexity and variety of the intergovernmental system that manages and supports our infrastructure."

In "Building the City Through the Back Door: The Politics of Debt, Law, and Public Infrastructure," David Perry provides a synthetic closing chapter that explores the centrality of public infrastructure policy to the institutional practice of American politics. In his chapter, Perry describes such policy as the product of a changing politics of constitutional/legal restrictions, fiscal relations, and public infrastructure systems. Therefore, to study public infrastructure is not only to study a key policy subject but also to study some of the key institutional relationships of domestic politics. Early in the chapter Perry makes this point in a different, more focused way: The public infrastructure process "includes services as proximate as today's potholes and as futuristic as tomorrow's information superhighway. It also triggers that most fundamental of questions in American politics: 'Who's going to pay?' . . . Building the public city," he goes on to suggest, "has always been as much an issue of finance as it has been an issue of technology transfer in the service of a social or economic need" (pp. 202-203).

The remainder of the chapter is an exploration of the intergovernmental history of this "issue of finance." The federal government from the very beginning has evidenced a conservative constitutional interpretation of its direct role in the public infrastructure process; Perry paints a rather fitful and uneven picture of federal participation in public works delivery—one that obtains even until today.

The absence of a concerted federal presence is mirrored by an active state role, especially in the first half of the nineteenth century, when states introduced a full range of debt-generating mechanisms in their highly active participation in public works policy. It is during this period that the practices of large-scale, unsecured, and highly speculative debt procedures were first introduced as key features of public infrastructure politics in the United States. With the fiscal demise of many of these fragile financial schemes, most states retreated from these decidedly unrepresentative debt practices, constitutionally limiting their own abilities to create debt by either imposing democratically acceptable ceilings or acquiescing to the wish of the voters. At the same time, states left the door wide open for local jurisdictions to engage in their own highly speculative and politically uncontrolled debt practices. During the latter half of the nineteenth century, the cities engaged in their own versions of what Perry calls the "backdoor politics of public works policy" as they became centers of rapid and sustained social and economic growth requiring significant public infrastructure delivery. By the time the states began to rein in the cities at the end of the century, they had amassed substantial amounts of unsecured and politically unauthorized debt in the service of public works delivery.

In the last half of the twentieth century this early intergovernmental tradition of passing on the bill of paying for public works from one level of government to another had been transformed to one of creating new tools of government (the public authority and various forms of revenue bonds) as new versions of fiscal "back doors" through which to escape constitutional restrictions on building the necessary public works systems of industrial and postindustrial urban America. The last sections of Perry's chapter describe the complex new array of fiscal and agential ways in which states and localities have continued to both constitutionally restrict borrowing for public works and simultaneously devise new means of circumventing these restrictions in order to keep paying for them. In short, Perry concludes that the key to understanding public infrastructure policy in the United States is understanding the practices of debt formation. Therefore, the future of public works policy in the United States rests squarely with those who practice the innovations and contradictions of debt formation; with those who continue to build the city through the back door.

■ Conclusion

The politics of public infrastructure have rarely been characterized by disagreements over the centrality of public works to the daily success of cities and the republic. Rather, as historian Seeley puts it "the issue has been how, not whether, to build, and more to the point, how to pay." As a result the politics of public infrastructure in this country have always blended the "how" of technology—how the train, auto, or fiber-optic cable can be employed—with the question of "how to pay"— how a private investment group, level of government, special district, or public authority can come up with the resources to, as the legendary public works operative Robert Moses used to put it, "get things done."

All the chapters in this book, the biographic study by Doig, the institutional analysis by Felbinger, and the stories of public works practice by Mier, coupled with the last three chapters on the governmental structure (Leigland), finance (Perry), and politics (Sanders) of building the public city, are concerned with *how* public infrastructure "gets done." Each of the chapters addresses the topic of "building the public city" as a multilayered discussion of history, political and administrative practice, governmental agency, and finance. Each essay is really a different take on the question of "How do public works get built?" And, finally, each essay presents a view of the process of building public infrastructure that is decidedly recursive, one where the building of public infrastructure is a highly contextual process, embedded in the ever-changing fabric of the nation and the city—where the infrastructure becomes both an ingredient of development and renewal and, in time, a condition or issue of change demanding development and renewal itself. Or, as Josef Konvitz, suggests: "Vital cities are never finished" (1985, p. 195).

NOTE

1. If there is a given in public infrastructure policy that cuts across all regions and levels of government, it is that there is inadequate effort, both fiscal and programmatic, attached to the maintenance and operations of public works. The reasons for such failures are multiple. First, few state and local politicians find much political currency in painting a bridge when they can build a new one. Second the federal government, driven by the same "monument-logic" of politics, contributes the lion's share of its funding to capital construction and little to maintenance. Most federally sponsored projects, once built, are turned over to state or local authorities for operation and maintenance. See in particular NCPWI (1988), Choate and Walter (1983), and Weiman (1993).

REFERENCES

Anderson, L. (1988). Fire and disease: The development of water supply systems in New England, 1970-1990. In J. A. Tarr & G. Dupuy (Eds.), *Technology and the rise of the networked city in Europe and America* (pp. 137-156). Philadelphia: Temple University Press.

Blake, H. (1984). *First interview on working with Robert Moses.* Buffalo: SUNY, Buffalo, The Robert Moses Research Project.

Castells, M. (1989). *The informational city: Information, technology, economic restructuring, and the urban-regional process.* Oxford, UK: Basil Blackwell.

Choate, P., & Walter, S. (1983). *America in ruins: The decaying infrastructure.* Durham, NC: Duke Press Paperbacks.

deHaven-Smith, L. (1985). Special districts: A structural approach to infrastructure finance and management. In J. Nicholas (Ed.), *The changing structure of infrastructure finance* (pp. 59-77). Cambridge, MA: Lincoln Institute of Land Policy.

Forester, T. (1987). *High tech society: The story of the information technology revolution.* Oxford, UK: Basil Blackwell.

Foster, M. S. (1981). *From streetcar to superhighway: American city planners and urban transportation, 1900-1940.* Philadelphia: Temple University Press.

Goodrich, C. (1960). *Government promotion of American canals and railroads, 1800-1890.* New York: Columbia University Press.

Johnson, S. (1992, April 19). River of our discontent. *Chicago Tribune,* p. 1.

Konvitz, J. (1985). *The urban millennium: The city-building process from the early middle ages to the present.* Carbondale: Southern Illinois University Press.

McKay, J. P. (1976). *Tramways and trolleys: The rise of urban mass transit in Europe.* Princeton, NJ: Princeton University Press.

Mier, R., with Giloth, R., Moe, K. J., Alpern, L., Harrison, B., McGary, H. M., Jr., Sherr, I., Vietorisz, T., & Wiewel, W. (1993). *Social justice and local development policy.* Newbury Park, CA: Sage.

Muschamp, H. (1994, February 13). Two for the roads: A vision of urban design. *The New York Times,* Arts and Leisure, Section 2, p. 1.

National Council on Public Works Improvement (NCPWI). (1988). Fragile foundations: A report on America's public works. *Final report to the President and Congress.* Washington, DC: Government Printing Office.

Peterson, G. E., & Miller, M. J. (1982). *Financing urban infrastructure: Policy options.* Washington, DC: The Urban Institute.

Porter, D., Lin, B.K.C., & Peiser, R. B. (1987). *Special districts: A useful technique for financing infrastructure.* Washington, DC: Urban Land Institute.

Rostow, W. W. (1960). *Stages of economic growth.* Cambridge, UK: Cambridge University Press.

Schultz, S. K. (1989). *Constructing urban culture: American cities and city planning, 1800-1920.* Philadelphia: Temple University Press.

Seeley, B. (1993, Winter). The saga of American infrastructure. *Wilson Quarterly,* pp. 21-39.

Tarr, J. A. (1988). Sewerage and the development of the networked city in the United States, 1850-1930. In J. A. Tarr & G. Dupuy (Eds.), *Technology and the rise of the networked city in Europe and America* (pp. 159-185). Philadelphia: Temple University Press.

Tarr, J. A., & Dupuy, G. (Eds.). (1988). *Technology and the rise of the networked city in Europe and America.* Philadelphia: Temple University Press.

U.S. Congress, Office of Technology Assessment (OTA). (1990). *Rebuilding the foundations: A special report on state and local public works financing and management.* Washington, DC: Government Printing Office.

U.S. Congress, Office of Technology Assessment (OTA). (1991). *Delivering the goods: Public works, technologies, management, and finance.* Washington, DC: Government Printing Office.

Warner, S. B. (1972). *The urban wilderness.* New York: Harper & Row.

Weiman, C. (1993, January). Road work ahead: How to solve the infrastructure crisis. *Technology Review, 97*(1), 43-48.

2

Politics and the Engineering Mind: O. H. Ammann and the Hidden Story of the George Washington Bridge

JAMESON W. DOIG

Othmar Ammann was at once a mathematician, a forerunner in the industrial revolution and a dreamer in steel. He was a master of suspension and a builder of the most beautiful architecture known to man, a combination of realist and artist rarely found in this highly practical world.

Robert Moses, *Poet in Steel* (1968)

O. H. Ammann is generally viewed as one of the great bridge builders of the modern world. He designed and then supervised the construction of most of the major water crossings completed in the New York region during the past 75 years—the Bayonne Bridge and the George Washington span during the 1920s and early 1930s, the Bronx-Whitestone in the

AUTHOR'S NOTE: An earlier version of this chapter was presented as a lecture at the annual meeting of the Swiss American Historical Society, on October 22, 1988, in New York City. The Society provided a grant in 1987 to further the research reported in this chapter, and additional funds were provided by the foundation Pro Helvetia and by the Alfred P. Sloan Foundation. The author is grateful for this financial support and for the research assistance of Alexis Faust, Janice Finney, and Paul Margie, whose activities were funded through these grants.

The author is also grateful for the encouragement, advice, and documentary materials provided by Margot Ammann Durrer, Werner Ammann, Sylva Brunner, and David Billington; and by Edward Cohen of Ammann & Whitney; Urs Widmer in Winterthur, Switzerland; Leon Katz and Vincent DeConzo at the Port Authority of New York and New Jersey; Karl Niederer and his staff at the New Jersey Archives; the staff of the New York State Archives; Elizabeth Lukach at the *Palisadian*; and Malcolm Borg and his associates at the *Bergen Record*.

Excerpts from *Poet in Steel* are reprinted by permission; © 1934, 1962 The New Yorker Magazine, Inc.

1930s, the Throgs Neck and Verrazano-Narrows bridges in the 1950s and 1960s. He had a crucial role in the design and building of the Golden Gate Bridge in the 1930s. And, as bridge engineer and chief engineer at the Port of New York Authority, he supervised the building of two other bridges—the Goethals and Outerbridge—as well as the construction of the first tube of the Lincoln Tunnel.

Ammann's bridges are important not only as engineering structures, but because of their impact on urban development in the New York area—for they have had a major influence in shaping the residential patterns, and the patterns of employment and recreation, across a vast bistate region.[1] As the quotation from Robert Moses suggests, Ammann's bridges also have an artistic quality that gives them, and their designer, a distinctive place in the long history of bridge building, and of civil engineering more generally, in the Western world.

Much has been written about Ammann—about his early life in Switzerland, his professional development in Europe and in the United States, and his engineering accomplishments. The articles and commentaries began as the great towers for the George Washington Bridge were being erected in the late 1920s; for the attention of engineers and journalists was drawn to this Hudson River crossing—which would be the longest single-span bridge in the world—and to its Swiss-American creator.[2] Ammann's artistic achievement was highlighted in a well-known essay by Le Corbusier in the 1930s: "The George Washington Bridge over the Hudson is the most beautiful bridge in the world. Made of cables and steel, it gleams in the sky like an arch upturned, blest. It is the only seat of grace in a disheveled city" (Le Corbusier, 1936, p. 75). During the past 5 decades the outflow of articles and papers has continued, and Ammann's ranking as one of the major figures of bridge design and engineering administration has been solidified.[3]

What is missing from the written record of Ammann's life and contributions is a crucial chapter on Othmar Ammann as "political entrepreneur"—on Ammann's major role in organizing public support for the first of these great bridges, the George Washington. This chapter in Ammann's life is critical to understanding why a bridge was constructed across the Hudson in that decade, permitting the rapid suburbanization of Bergen County and the surrounding hinterland. It is also crucial to explaining why a young engineer of only moderate reputation was chosen to design and build that gigantic span, and how it came to pass that this engineer was able to use his technical and artistic talents to create the series of great bridges that stamped him as a major figure in the engineering profession. Finally, Ammann's political activities are significant to an understanding of why the Port of New York Authority left the field of railroad planning to become a leading player in the new age of rubber-tired transportation.

The purpose of this chapter is to sketch out an answer to these questions, and in doing so to illustrate two broader points: (a) that biographical analysis can be valuable as a route to increasing our understanding of patterns of political power and social causation in a society; and (b) that close biographical probing may be especially helpful when previous writings have treated an individual as an example of an "ideal type" of his or her specialized profession or occupation.

The chapter is divided into several parts. The first section provides a brief summary of Ammann's life and work, as they have been described in a dozen biographical essays and magazine articles, a book-length biography, and commentaries on Ammann in many books—on bridge engineering, on the development of the New York region, and on the history of the George Washington Bridge.[4]

In these published accounts, there are some puzzling aspects and apparent gaps. The second part of the chapter identifies these and briefly notes how the events of the "missing years" in Ammann's early career were uncovered.

Then I turn to those missing years and summarize the steps through which Othmar Ammann confronted a contentious and often hostile environment, found his early hopes for a major span across the Hudson shattered, sketched out his own approach to a Hudson bridge, and then entered the ranks of the unemployed—where he began, haltingly, to put together a coalition that would provide political support for a Hudson River bridge, and that might indeed endorse his own vision of such a bridge, placed where *he* thought it should go. It is his efforts in building this coalition, and in orchestrating its activities toward a successful conclusion, that justify calling Ammann, the engineer and artist, by the additional title of political entrepreneur.[5]

The final sections of the chapter consider the broader questions listed above.

■ The Life of a Great Bridge Engineer: The Standard Account

> Keen instruments, strung to a vast precision
> Bind town to town and dream to ticking dream.
> *Hart Crane, "The Bridge" (1966)*

Ammann was born in 1879 in the canton of Schaffhausen, in Switzerland. In 1898 he entered the ETH Zurich (the Swiss Federal Polytechnic Institute), where he studied with Wilhelm Ritter, a distinguished bridge designer. Graduating in 1902, he worked as a structural draftsman in Europe for 2 years and then, at the urging of one of his former professors, left for America.[6]

Ammann arrived in New York in the spring of 1904 and found employment with a local engineering company. During the next several years he worked for engineering firms in Manhattan, Chicago, and Pennsylvania, and he worked on several major bridges, including the Queensboro in New York City. The possibility of a bridge across the Hudson also attracted his early attention.[7] In 1912 Ammann joined the firm of Gustav Lindenthal, a railroad-bridge engineer with an international reputation, and Lindenthal soon appointed him as his chief aide in work on the Hell Gate Bridge.[8] Much of Ammann's time during the years 1912-1917 was devoted to the Hell Gate, where he was in charge of all office and field operations, supervising a team of 95 engineers (see Cohen, Stahl, & Wilson, 1987, p. 8).

The Hell Gate Bridge was completed in early 1917, and Lindenthal then had very little engineering work to occupy his staff. He suggested that Ammann take a temporary position in New Jersey, managing a clay pottery mine in which Lindenthal had invested. Ammann took that job, at the Such Clay Pottery Company in Middlesex County; and with his managerial skills he turned a shaky financial enterprise into a healthy firm, which earned a modest profit for Lindenthal and other investors. In 1920 Lindenthal called him back to assist in developing plans for a gigantic railroad-vehicular bridge that would cross the Hudson at 57th Street.[9]

Ammann worked on this project with Lindenthal from 1920 until 1923. By the middle of 1922, however, Ammann had become concerned that the 57th Street Bridge could not be constructed in the near future, because of its great cost, the reluctance of the railroads to commit themselves to using the bridge, and the opposition of Manhattan business and political leaders to a project that would dump 20 lanes of traffic into the midtown area. Ammann urged Lindenthal to cut down the size of his project and to shift its location north of midtown Manhattan. Lindenthal resisted, and in the spring of 1923 Ammann left his employ and entered private practice on his own (Cohen et al., 1987, p. 9; Widmer, 1979a, p. 11).

After 2 years in private practice, Ammann was hired by the Port of New York Authority in July 1925 as its Bridge Engineer, and he was placed in charge of the design and execution of the proposed span across the Hudson—between Fort Lee in North Jersey and 179th Street in Manhattan. He was also given supervisory control over construction of three smaller Port Authority bridges, between New Jersey and Staten Island.

All of the biographical and other accounts agree that the Port Authority's decision in 1925 to commit its energies to constructing a bridge at 179th Street, and to hire Ammann to carry out that challenging task, were the crucial steps in the flowering of his career.[10]

Ammann stayed with the Port Authority, as Bridge Engineer and then as Chief Engineer, until 1939, with collateral duty in the 1930s as chief engineer for Robert Moses's Triborough Authority. During these years, he designed and constructed the George Washington, the Bayonne, and the Bronx-Whitestone bridges; he also supervised construction of the Goethals, Outerbridge, and Triborough spans in the New York region, and he was an influential adviser in the designing of the Golden Gate Bridge. By 1939, however, the Great Depression had taken its toll on the financial health of the Port Authority, and on its energy and vision. There were no challenging projects in the offing, and Ammann then left the agency, joining forces with another engineer to found the firm of Ammann & Whitney, which in the next several decades worked on a wide range of projects around the world.

During the 1940s and early 1950s, Ammann's old employers—the Port Authority and Robert Moses—engaged in intermittent warfare over control of airports and vehicular projects in the New York region, but by the mid-1950s they agreed to combine their energies on behalf of several new highway projects. Once again they called upon Ammann— to design and supervise construction of a new bridge across the East River (at Throgs Neck), a lower deck for the George Washington Bridge, and a structure that would again give to New York the longest single-span bridge in the world, the Verrazano-Narrows Bridge across the entrance to New York harbor.[11]

The Verrazano Bridge was completed in 1964, and the next year, at the age of 86, Othmar Ammann died. On the centennial of his birth, in 1979, celebrations were held in New York and in Switzerland, and the Swiss government issued a stamp in honor of this leading citizen of both countries.

■ **The Puzzle and a Search
    for the Missing Pieces**

> Luck is important to those who work in the structural arts; great engineer-
> ing assignments are comparatively rare, and it takes great assignments to
> make great engineers. In that respect, Ammann has been lucky....
>
>   M. MacKaye, *"Poet in Steel,"* The New Yorker *(1934)*

The paragraphs above summarize the well-known biographical account. But there are some gaps or puzzling aspects of the story. For
example, why did the Port Authority hire Ammann in 1925 to take
charge of designing and constructing a giant bridge across the Hudson?
Could "luck" possibly explain the happy choice? In view of the size and
regional importance of this project, prominent engineers around the
world might be expected to seek the commission; and because of the
importance of the project to the Port Authority (in 1925 it had no
operating facilities at all), that agency might have been expected to
reach out for a bridge engineer with a major reputation and a record of
independent accomplishment. Ammann had neither; he had been second
in command to Lindenthal on the Hell Gate and other, smaller projects,
and he had won a prize for a published paper on the Hell Gate Bridge.
He had no significant engineering achievements to his own credit.

Ammann's limited reputation had in fact cost him a commission at
the Port of New York Authority only a few months earlier. In the fall
of 1924 he had submitted a bid to design two smaller bridges for the
Authority, but his offer had not been accepted; the agency concluded
that for their first operating projects, they would need to turn to "an
engineer of long established reputation."[12]

Related to the question of why the Port Authority chose Ammann is
a second issue: Why did the port agency—which had been created to
carry out a plan for better railroad connections and rail freight terminals
in the New York region—decide to build a gigantic bridge for motor
vehicles?

Published materials provide important elements of an answer, but
they leave significant gaps as well. Part of the answer lies in the attitude
of the railroads, and in the growing importance of motor trucks. During
the 1920s the Port Authority's efforts to improve outmoded rail facilities made little progress, primarily because of resistance from the
region's dozen rail corporations. Meanwhile, freight distributors expanded their use of over-the-road vehicles for goods shipments within

the New York region and beyond; but efficient truck movement across the region was stymied by the Hudson River, which had no tunnel or bridge crossings.[13] To overcome the Hudson barrier, New York State and New Jersey had created joint state Bridge and Tunnel Commissions, which in 1920 began a tunnel under the Hudson, between Jersey City and Canal Street in Manhattan. During the next 2 years, the "cooperating" commissions disagreed about engineering and financing issues, and progress on the tunnel was delayed. After 1922, however, the joint effort proceeded harmoniously, and commission members began to think about constructing a *series* of vehicular tunnels—possibly at 40th Street in Manhattan, at 125th Street, and several in between. Private investors began to gather support for a tunnel crossing, too, perhaps at 125th Street.[14]

At this point, Alfred E. Smith, who was elected governor of New York in 1922, entered the picture. Smith did not believe that private corporations should build and control major highway arteries across the Hudson, and in early 1923 he vetoed a bill that would have authorized such ventures. He was also skeptical of relying on the existing interstate commissions, which had shown a proclivity for internecine warfare, and which would need state funds to construct a series of interstate tunnels. Smith's clear preference was to place the development of all interstate crossings for motor vehicles as well as railroads entirely in the hands of the Port of New York Authority, which was authorized to use toll revenues to pay for its own projects—potentially sparing Al Smith and other state officials from the burden of using tax revenues for bridge and tunnel projects.[15]

Smith's views were important, but they could not be determinative. The Port Authority could assume these wider duties only if both New York and New Jersey passed new legislation, and there was strong sentiment in the Republican-dominated houses in both states to rely on the joint tunnel commissions or on private ventures. New Jersey's Governor Silzer also sought to interest private capital, and through most of 1923 and perhaps well into 1924, the Port Authority's own staff did not appear much interested in adding vehicular bridges to its duties.[16] Throughout much of 1923 one might reasonably have predicted that the Canal Street tunnel would be followed by a series of other tunnels under the Hudson—perhaps next at 40th Street and 110th Street—built by the joint commissions, and possibly with one or two financed by private investors.

However, in 1924 local civic groups and public officials in North Jersey began to demand that the Port Authority construct a great bridge at Fort Lee. An influential Republican senator from that area championed the cause and introduced legislation, and Governor Silzer joined the public campaign, urging that the Port Authority take action. In New York, local business groups took up the cry. Early in 1925 both state legislatures passed legislation that compelled the Port Authority—which may still have been reluctant to take on the task—to turn its energies to designing a Hudson River bridge.

To restate the second question, therefore, in altered form: Why was the divided sentiment on *where* to cross the Hudson, and *how* (bridge versus tunnel; joint commission versus Port Authority versus private initiatives) replaced during 1924 by a clear, sustained demand for Port Authority action to build a great bridge at 179th Street?

The published biographical record on O. H. Ammann suggests a third question as well: What was Ammann doing between the time he left Lindenthal in the spring of 1923 and the summer of 1925, when he joined the staff of the Port Authority? The published materials indicate that he was in private practice as an engineer, and that he investigated various sites for a trans-Hudson vehicular bridge and decided a span between 179th Street and Fort Lee would be the best location. But beyond that sparse description, the published record provided little information. As it would turn out, the detailed answer to this third question would also provide major clues to answering the first two queries.

In the spring of 1987 my primary motivation in thinking about these issues was not biographical, with a focus on Othmar Ammann. Rather, it was historical and political, as I tried to sort out the forces that led the Port Authority to change direction in the 1920s, leaving the field of rail freight and embracing the automotive age—a field in which it would become famous, or infamous, depending upon one's point of view. It seemed evident that the decision to hire Ammann and to put him in charge of the Hudson bridge project signaled the Authority's commitment to this new direction. To determine why the Port Authority hired Ammann might cast light on the larger issue.

Unfortunately, though the published literature on the Port Authority's early history and on the politics surrounding the building of the George Washington Bridge was quite extensive, it was not very helpful on this point. Ammann seemed to appear, essentially, from nowhere.[17]

One possible research strategy at this point, and the one I adopted, was to temporarily set aside the wealth of materials on institutional history, interest group pressures, and the dynamics of interstate conflict, and pursue the issue biographically—looking for information on Ammann's earlier career that might help to clarify the political issue of direct concern.

Where to turn for answers? A search of the Port Authority's files revealed no information on Ammann's early career or why he was chosen.[18] However, David Billington, professor of civil engineering at Princeton, did have some materials on Ammann, including one booklet with a section titled "Autobiography"—which contained four pages of notes in Ammann's handwriting, listing brief information on his birth and education and on his activities from the late nineteenth century through 1956.[19]

Ammann never fleshed out these sketchy notes, but a few lines provided useful clues in answering the several questions noted above:

Own studies & promotion activities for bridging Hudson at 179th Str. Promotion of G.W. Br;

Rea — Silzer Dwight Morrow Binder

Creation of the Port of N.Y. Auth. in 1921? — its primary purpose — incidental functions of

financing & building of bridges (Arthur Kill Br. — G.W. Br. — Bayonne Br. Later

Holland T. — Lincoln T. P.A. Bldg. Commissioners & Manager

Chf. Eng of Br. — Chf Eng.

The first line indicated that Ammann was active in promoting the idea of a bridge at 179th Street, but it was not clear whether this was before or after the Port Authority was created in 1921. The reference to "Silzer" in the second line suggested that Ammann had had some contact with Governor Silzer during this period, and a search for biographical information on Silzer yielded an early connection between the two: Before his election as governor in 1922, Silzer had been a political leader in Middlesex County and a member of the board of directors of a local business, the Such Pottery Company, where Ammann was the operating manager in 1917-1920.

The next step was to look at the official gubernatorial papers, stored in the State Archives in Trenton. An hour with those materials suggested that the Silzer files would go a long way toward answering all three of my questions.[20] The files contained more than 100 letters, drawings, and notes, most of them between January 1923 and July 1925. The great bulk of the materials was correspondence between Silzer and Ammann, but there were also letters between the governor and Port Authority officials, the governor and Gustav Lindenthal, and the governor and Dwight Morrow, a member of the J. P. Morgan banking firm, who provided advice to Silzer and Ammann on the possibility of private financing for a bridge across the Hudson.

There were other leads to pursue. Billington had suggested that Edward Cohen, currently the managing partner at Ammann & Whitney, might be helpful. Cohen sent a paper he had written, and he suggested that I call Othmar's daughter, Margot, and his son, Werner. Werner had some recollections of the early 1920s, but he had been away at college during part of the missing years, and he had no detailed knowledge of his father's activities on the Hudson Bridge project. Margot had fewer recollections of the early 1920s (she was 1 year old in 1923), but she did have a most valuable source—letters from Ammann to his mother in Switzerland, written between his arrival in the United States in 1904 and her death in 1928. These were in German, and she offered to translate them, an offer I gladly accepted.

Margot Ammann and Edward Cohen also suggested another important source of information: the Ammann archives in Winterthur, Switzerland, which are maintained there, together with an exhibition on his works, by Urs Widmer, mayor of the city and at one time a staff member at Ammann & Whitney. During the summer of 1988 Widmer went through Ammann's diaries and other materials on file in Winterthur and sent me dozens of pages of diary entries and other information, which helped to fill in the story of Ammann's missing years.

To confirm and amplify Ammann and Silzer's descriptions of the bridge campaign, it would also be desirable to locate newspapers published during the 1920s that carried articles on the organizing activities of Ammann and others concerned with Hudson River crossings. The most detailed accounts appeared in the *Palisadian*, a weekly newspaper published in a town near the New Jersey terminus of the George Washington Bridge. The *Palisadian* has appeared continuously since 1906. Its reporters gave extensive coverage to the campaign for the bridge, and—unlike the old issues of many weeklies in the state—the editions had been bound every year and were stored at its offices. More than two dozen articles on the activities of Ammann and other supporters of a vehicular bridge in that area were published during the years 1923-1925.

■ The Team of Ammann and Silzer

> To take a stand, to be passionate . . . is the politician's element, and above all the element of the political leader.
>
> *Max Weber, "Politics as a Vocation" (1919)*

What emerges from the records in the New Jersey State Archives—supplemented by Ammann's letters to his mother, his diary entries, and newspaper articles—is the creation of an informal but close alliance between Ammann and Silzer, beginning in the winter of 1922-1923. Ammann was certain that the technology was available to permit a single-span bridge to be cast across the Hudson in the vicinity of 179th Street in Manhattan—even though such a span would be nearly *twice* the length of any bridge span yet constructed.[21] And the new technologies should make it possible to meet high standards of esthetic as well as technical achievement—a combination that might attract any first-class engineer. Moreover, the expanding use of automobiles and trucks across the New York region—exemplified by long lines of vehicles waiting to use trans-Hudson ferries from Bergen and northern Manhattan—indicated to Ammann that there was a real need for a bridge located well north of congested mid-Manhattan. Therefore Ammann was ready to throw his considerable energies into the dual task of working out the detailed design requirements needed to show that such a bridge was technically feasible, and helping to create the public support needed before the great span could be approved and constructed.

For his part, Silzer could see the economic advantages to New Jersey that would flow from improved transportation between New York City and the North Jersey suburban areas, and he could see the advantage to his own political prospects that might follow if the great bridge, with its stimulus to commerce across the northern part of the state, were commenced during his term in office. However, Silzer was a Democrat, and the area of his state that would be most directly affected by a new bridge at Ammann's preferred location was Bergen County, a major Republican stronghold. Silzer and his party were not much loved in that northern suburban area, nor in the state legislature, which was also controlled by the party of Lincoln. If Silzer hitched his political star in a public and sustained way to a campaign to build a great Hudson bridge, neither would be likely to benefit. It would be better for Silzer to leave the visible organizing efforts to local interests in North Jersey and New York, which would probably be ready to campaign vigorously once Ammann had demonstrated the engineering and economic feasibility of the great design.[22]

During the next 2 years the campaign went forward, with Ammann frequently spending his days in the political trenches, while at night he was bent over engineering drawings and calculations.[23] Silzer was absorbed largely in other policy issues, accompanied by recurring political battles with an unfriendly state legislature, but he provided constant encouragement to Ammann and occasional guidance on political strategies. At several important points he also intervened directly to promote Ammann's interests, which he made essentially his own.

Before describing the nature of that alliance and Ammann's activities in more detail, I should say a little more about related developments in the years just prior to and during the winter of 1922-1923.

Two Engineering Mentalities

The divisions and conflicts between Ammann and Gustav Lindenthal—Ammann's mentor, benefactor, and boss for many years—are crucial to the story of the George Washington Bridge. And they illustrate two very different relationships between "politics and the engineering mind."

Both Lindenthal and Ammann were civil engineers on a grand scale—bridge builders who came from Europe to the United States because the young nation had the vast expanse, the wide rivers and deep ravines, that could provide great challenges to engineering designs and construction skill for years to come. Moreover, America had the commercial vitality and urge for efficiency in transport that would require that great bridges be cast across the East River, the Hudson, the Mississippi, and other waterways; and she had the wealth that would allow resources—great amounts of manpower and materiel—to be gathered and orchestrated and used according to the designs of great engineers. And, more generally, by the late nineteenth century the nation that had built the Erie Canal and the transcontinental railway seemed imbued with a philosophy about planning and building that fit the hopes and aspirations of these two engineers and their compatriots. The planner Daniel Burnham expressed the basic American value that underlay the great projects and attracted Lindenthal and Ammann from their home countries:

Make no little plans; they have no magic to stir men's blood. . . . Make big plans; aim high in hope and work, remembering that a noble logical diagram once recorded will never die. . . . Remember that our sons and grandsons are going to do things that would stagger us. Let your watchword be order and your beacon beauty.[24]

Lindenthal got here first. Born in Austria in 1850, Lindenthal studied engineering in Europe and crossed the Atlantic in 1874. After working for several years in Pittsburgh, he came to New York and drew up his first plan for a bridge across the Hudson in 1888. The span would rise on the Jersey shore at Hoboken, land at 23rd Street in Manhattan, and carry trains over the river on 10 railroad tracks.[25] The Pennsylvania Railroad, which had to transfer its thousands of rail passengers each day to ferries crossing the Hudson, was strongly interested, but the project was delayed as the Pennsylvania tried to work out a joint plan with other rail lines that also deposited their Manhattan-bound passengers at ferry terminals on the Jersey shore.

Meanwhile, Lindenthal was appointed Bridge Commissioner for New York City in 1902, where he completed the Williamsburg Bridge and planned the Manhattan and the Queensboro bridges (all three spanning the East River, from Manhattan to Brooklyn or Queens). He found working in the political environment of New York City to be personally and professionally frustrating, and in 1904, after a series of conflicts with city officials on engineering and esthetic issues, he resigned.[26] Thereafter Lindenthal devoted his energies mainly to privately sponsored enterprises. His most important project after 1903 was the Hell Gate Bridge, commissioned by the Pennsylvania Railroad, which he began designing in 1907 and completed in 1917. It was the largest single-span arch bridge in the world.[27]

With the Hell Gate project completed, Lindenthal turned his attention once again to the great bridge across the Hudson. Now he moved it uptown, to 57th Street; noting the increasing importance of automobiles and trucks, he added 20 lanes for vehicular traffic. The entire structure would cost $100 million, Lindenthal estimated, but he was certain that private capital could be raised to meet this total, and that the Hudson bridge would be a profitable undertaking—as well as an immense benefit to North Jersey and New York.

As Lindenthal prepared to develop the designs needed for the 57th Street bridge, and to begin a campaign to raise the necessary money and political support, he turned to Ammann for assistance.[28] As noted earlier, Ammann had been one of his top aides on the Hell Gate during the years 1912-1917; during the years 1917-1920, Ammann had been manager of the Such Clay Pottery Company, working under the board of directors, which included both Lindenthal and George Silzer.

In 1920 Ammann rejoined Lindenthal in New York, becoming his principal assistant at the North River Bridge Company—the corporate vehicle, created by Lindenthal, through which he hoped to obtain funding for the 57th Street span as a private bridge, and the organization that would build the huge structure. For 2 years, Ammann worked loyally and energetically on engineering issues associated with the plan. And at first, he worked with considerable enthusiasm, judging from his report home:

> The new project brings me great satisfaction, it is a great noble structure, and . . . the concept and modeling of the project demand intense attention and work. . . . It will be possible for one-half million passengers and 12,000 vehicles and 4000 tons of freight to pass over it per hour. . . . The towers will be as high as the tallest skyscraper in New York.[29]

During these 2 years, however, Lindenthal's plan received a series of debilitating blows: Influential civic and business interests announced their opposition to a bridge at 57th Street, fearing that it would make traffic congestion in the midtown area intolerable; the bistate Port of New York Authority rejected it for the same reasons, and was pressing ahead with an alternative plan, involving rail tunnels under the Hudson and under New York Bay; and the major railroads and other potential investors refused to invest the millions needed to make the project a reality. By the fall of 1922, Lindenthal's financial situation was very rocky, and his company could only afford to pay Ammann part of his monthly wages.

At this point, the close alliance between the two men began to sunder; and it soon became clear that their underlying values and worldviews were quite different. Both men shared the enthusiasm for the great project that Lindenthal had created in his mind and on paper. But to Lindenthal, there was no alternative; neither tunnels under the Hudson, nor bridges farther upriver, nor a bridge limited to motor vehicles (and therefore much cheaper than a bridge strengthened to carry railroad trains)—none of these could possibly meet the vast need for improved transportation between Manhattan and the western regions. Moreover, as New York City's bridge commissioner, Lindenthal had had a taste of engineering work in a highly political environment, and he was not inclined to rely on that uncertain and conflict-filled route. The rational working environment, the corporate sponsorship, the freedom to design, and the international acclaim accompanying his Hell Gate project had demonstrated the right way to do it. In Lindenthal's mind, insulation and professional integrity were crucial, and the engineer could be relied upon to work out the best way to meet the other goals and constraints within which all great engineering projects must be designed and carried out—concerns with economy and with the esthetic quality of the project itself, and an understanding of how the project would fit into the broader patterns of economic and social relationships within the region.[30]

Ammann's perspective was different. Lindenthal's great bridge, in its original dimensions, would be wonderful. To Ammann, however, the substantive arguments and the political strength of the opponents deserved the same steely-eyed analysis that a good engineer devoted to understanding the stresses on bridge cables and the stability of the ground under proposed bridge towers. If influential opponents thought the bridge too large, or badly located, Ammann could draw on his engineering experience and perhaps find ways to modify the plan, rather than let it go down to defeat. And as to the financing problem, private funding would be welcomed—because that would reduce the prospect that the project would become mired in the conflicts of politics. But if private investors could not be attracted, then a great bridge sponsored by government was better than no Hudson crossing at all—or the awful alternative (to a bridge engineer) of tunnels in the mud, under the Hudson.[31]

On the subordinate issues of location, type of bridge, sponsorship and financing, Ammann was a pragmatist. His main goal was to span the Hudson. Of course he had the psychological advantage, in thinking about alternatives, that the original design was not *his* plan; Lindenthal had been working on a railroad bridge across the Hudson to the middle of Manhattan since 1888—for more than 30 years. And Ammann was a mere 42, nearly 30 years younger than his boss; possibly youth permitted greater flexibility.Perhaps most important, however, Ammann—in contrast to Lindenthal—could encompass political obstacles, and strategies to overcome them, within the analytic framework of his engineering mind. Any good engineer knew, for example, that you had to design your bridge in relation to the character of the terrain where the towers would sit. Therefore, if preliminary studies suggested the tower footing would be solid rock, and closer exploration revealed softer ground, adjustments and even major redesigns would be necessary; and sometimes long weeks and months of arduous work would be needed to solve the problem and ensure that the tower and the bridge would hold. Moreover, bridge engineering was not an armchair activity; you had to go into the field continuously, marshal and motivate your workers, and modify your abstract designs as the land and the weather and the impact of human mischance required.

So too, close exploration of the *political* ground associated with any large project was essential; and this exploration might require meeting with local politicians and business people and interested citizens, in towns and county courthouses across the region, in order to work through the proper combination of engineering, esthetic, and political designs.

Or does this stretch the concept of an "engineering mind" too far? Perhaps Ammann can simply be viewed as a first-class engineer who, for a brief time and when it was absolutely necessary, showed that he also had the separate talents of the first-class political entrepreneur. My guess is that the talents were more closely joined, at least in this taciturn but passionate Swiss-American.

When Ammann pondered the problems that confronted Lindenthal and the 57th Street bridge in the fall of 1922, he concluded that the best way to meet these difficulties was to reduce the size and cost of the proposed bridge. If the bridge were limited to motor vehicles and light rail transit, the heavy, expensive structure needed for freight trains could be replaced by a lighter span, at much lower cost—and private investors might well be attracted to invest in a moderate-cost toll bridge for vehicles. Also, if influential citizens opposed a bridge at 57th Street, why not construct a bridge farther north, away from the congested midtown area? Once that crossing proved successful, a modest bridge at 57th Street might also receive wide support.

During the fall and winter of 1922, Ammann urged Lindenthal to reduce the size of the 57th Street project, and to consider shifting his short-term goal to an uptown bridge for autos, trucks, and light transit. But Lindenthal thought a crossing uptown would be too far north to attract much vehicular traffic; also, railroads as well as motor vehicles needed better access to New York, and a railroad bridge would bring freight and passengers to mid-Manhattan with marvelous efficiency. To Lindenthal, the 57th Street project was the only satisfactory way to solve the region's major freight and passenger problem—how to integrate New York City's vast economic strength and its people with the economies of New Jersey and the rest of the continent.

As 1922 drew to a close, it seemed evident to Ammann that he would have to take some initiative on his own. His diaries and other writings during these months do not provide clear evidence that he was deeply distressed, and his letters to his relatives are guarded. But a year later Ammann expressed his feelings on Lindenthal, the 57th Street bridge, and the possibilities for a brighter future, in a frank letter to his mother:

> In order for you to understand my situation for many months, in fact for a whole year, I will no longer conceal from you that the giant project for which I have been sacrificing time and money for the past three years, today lies in ruins. In vain, I as well as others have been fighting against the unlimited ambition of a genius who is obsessed with illusions of grandeur. He has the power in his hands and refuses to bring moderation into his gigantic plans. Instead, his illusions lead him to enlarge his plans more and more, until he has reached the unheard of sum of half a billion dollars—an impossibility even in America.
>
> However, I have gained a rich experience and have decided to build anew on the ruins with fresh hope and courage—and, at that, on my own initiative and with my own plans, on a more moderate scale. It is a hard battle. . . .[32]

But if Ammann felt compelled to strike out on a new course at the end of 1922, perhaps breaking free of Lindenthal, he would need to forge a new alliance to help him achieve his goal. And this brought him to George S. Silzer.

Silzer's Aspirations

In November 1922, George Silzer, Democrat from Middlesex County and a former state senator, was elected Governor of New Jersey. He would serve as the state's chief executive from January 1923 until January 1926, with the state legislature controlled by the Republicans throughout those years.

Silzer was an activist in his philosophy of government, a Wilsonian Democrat. Indeed Silzer had been one of Woodrow Wilson's chief aides in the legislature when Wilson was governor in 1911-1913 (Link, 1947, pp. 245-246). One of Silzer's strongest interests before and during his years as governor was the improvement of highway transportation; he viewed this goal as crucial to the state's economic growth. Thus he supported extensive road-building programs, as well as bridges that would connect his own Middlesex County to nearby Staten Island.[33]

Consistent with his reputation as a Wilson Democrat, Silzer denounced the logrolling methods and inefficiencies of the existing county and state highway agencies; he created a new state highway body to devise an efficient road system, and he urged "scientific planning" in all areas of state government.

The possibility of opening the large rural areas of northeast Jersey to rapid economic development was attractive to Silzer; and a crucial step in achieving this goal would be improved access between the northern counties and New York City. So the new governor might be in a receptive mood if Othmar Ammann could bring him a feasible plan to dissolve the Hudson barrier.

The Alliance

Silzer's victory in the 1922 election provided Ammann with an opportunity to reach out for assistance, as he tried to extricate himself from Lindenthal's fixation. Ammann, Lindenthal, and Silzer were already well acquainted through their mutual interest in the Such Clay Pottery Company.[34] Moreover, it seems clear that Silzer and Lindenthal had discussed the 57th Street bridge.[35]

At some point in the weeks before George Silzer took office in mid-January, Ammann talked with him about the need for a Hudson River crossing, about the economic and political problems that surrounded the Lindenthal bridge, and about the advantages of a bridge farther north, joining vast and rural Bergen County to the urbanized eastern shore. The bridge could be limited to motor vehicles and light trolleys, which meant that its cost would be only one quarter of the Lindenthal colossus. Ammann had studied various sites, and he preferred a bridge that swept from the Palisades cliffs, in the town of Fort Lee, across the River to 179th Street in Manhattan.[36]

We have no direct account of the views Silzer expressed at this meeting, but later evidence (discussed below) clearly indicates that he was enthusiastic about Ammann's proposed bridge, and that he hoped it might be financed by private capital. Moreover, Silzer was wary of the alternative "low cost" way to overcome the Hudson barrier—a series of tunnels under the Hudson.[37]

But if a Fort Lee bridge was a promising idea, what role should Silzer take in advancing the cause? Bergen County was Republican territory, and Democrat Silzer apparently felt it would be unwise to associate his name too closely with a crossing that would need active Republican support if it were to be approved by the state legislature; if it were viewed as "Silzer's bridge," Ammann's proposal might fail. However, Silzer could offer guidance on how Ammann and others interested in the Fort Lee bridge might gain public support; he could contact financial people confidentially, in the hope that private capital might be attracted to the scheme; and he could approach the recently created Port of New York Authority to ask if it would endorse such a bridge as consistent with its general goals.[38]

The major burden, then, fell to Ammann, and as the new year opened he faced a difficult task. He would need to develop local interest in his Fort Lee bridge—within the communities of Bergen County and nearby Passaic County, and across the river, in northern Manhattan and the Bronx, and perhaps further north in Westchester and southwestern Connecticut. Indeed, he would need to persuade local business leaders and elected officials not only that his bridge was an interesting idea, but also that it was the *best solution* for the near future—better for the communities and for the economic growth of the region than the alternatives already being actively discussed. Two of these alternatives— underwater tunnels at 110th Street and at 125th Street—also promised economic benefits for Bergen and Passaic, and for New York City; moreover, a private association had already been created to press for the 125th Street tunnel, and editors at Bergen's major daily newspaper, the *Bergen Record*, were supporting that plan.

Then there was the 57th Street bridge. Ammann had not yet broken with Lindenthal; and he still hoped the great man might be persuaded to join forces with Ammann—perhaps take the leading role—and attract the private capital and political support needed for the Fort Lee plan. But if Lindenthal stood firm, the prominent civic leaders and financial men who had joined his board of directors would probably stand with him; and although that combined force might never produce a real bridge at 57th Street, its opposition might doom Ammann's uptown scheme.

How best to proceed? For Ammann, the answer came in three parts: He would need to sketch out a bridge design that was so dramatic, so arresting, that it would claim the attention and approval of the attentive publics of North Jersey and New York. And he would need to work out the probable cost for the bridge, so that it would strike a chord as financially feasible—in contrast to Lindenthal's gigantic scheme. Then he would need to take this design, and his ideas on how the bridge would benefit the region, directly to the public officials and local groups on both sides of the Hudson.

The first two steps were, for Ammann, comparatively easy. Trained in the great Swiss tradition, and apprenticed with Lindenthal, Ammann's developed engineering skills were a match for his considerable esthetic instincts. And he had had many years' experience, with Lindenthal and earlier, in working out the detailed costs associated with bridges large and small.[39] Beyond this training and experience, Ammann had the creative insight that bridges of longer span might carry sufficient deadweight to make extensive vertical trusses unnecessary. Ammann's own careful analysis supported this intuition, and he was then able to design a bridge which was strikingly light in appearance, and which carried a cost estimate proportionately lower than other long-span bridges.[40]

The third step—knocking on doors, trying to convince skeptical or preoccupied local officials, newspaper reporters, and shopkeepers— that was a different story. Ammann had confidence in his professional abilities, and in the value of a great bridge at Fort Lee. But he was a modest man, and one who did not talk easily of his interests and his passions, especially when those interests would be linked to advancing his own career.[41] However, if he were to make any headway in developing public support for a bridge between Fort Lee and 179th Street, Ammann would have to break through his natural reticence and advocate action on his plan—until a civic organization could be formed to take the leading role in this public relations effort.

This third step would also mean that Ammann would have to break formally with Lindenthal—unless he could convince Lindenthal to join him—so that Ammann could campaign openly for the uptown bridge. In the short run, this break would almost certainly mean that Ammann would have to join the ranks of the unemployed: To carry out the engineering studies for the Fort Lee crossing, and to campaign for approval, would absorb almost all his waking hours. There would be no time available to work on other projects with another engineering firm.

Finally, before the campaign could be successful, he and Governor Silzer would have to find an operating organization that could take Ammann's designs, raise the funds needed, and actually build the bridge. If Lindenthal were to change his views, the great man's North River Bridge Company could do the job. Otherwise, Ammann and Silzer would need to explore ways of creating a separate private corporation, or perhaps consider what kind of governmental agency might undertake the complex project.

■ Into the Political Arena

In the first months of 1923, Ammann and the governor began their joint campaign. On January 9, Ammann reported to Silzer that he had met with the governing board of Bergen County, and that their initial reaction was to support "the bridge at Fort Lee."[42] They also agreed, Ammann said, with Silzer's view that no new vehicular tunnels under the Hudson should be constructed until the Holland Tunnel was in operation.

A week later Silzer was sworn in as governor, and in his inaugural address, he referred to the advantages of northern New Jersey: "It is especially attractive to those who find the congestion of New York City unbearable, and who seek to live in a section at once high, healthy and accessible" (Silzer, 1923). But North Jersey was not really accessible, Silzer pointed out, especially to motor vehicles, which had to wait for hours to cross the Hudson by ferry. It was now time, Silzer argued, to give close consideration to building a bridge across the Hudson, a bridge "of ample size to care for vehicular and passenger travel, and for railroad terminal service."[43]

During the spring of 1923 Ammann met with several local groups in North Jersey, described his idea for a wide span at Fort Lee, and received some encouragement—but no one offered to take on the major task of organizing support across the counties that would benefit from the bridge. Meanwhile, Silzer contacted Dwight Morrow, a Wall Street expert in finance, for an evaluation of the prospects that Lindenthal's bridge could be built with private capital; Morrow's response, on March 2, was decidedly pessimistic. And Ammann, noting that opposition by Manhattan interests had not abated, and that the large cost of the 57th Street bridge was a major obstacle, urged Lindenthal to cut down the size of the bridge and move it northward. But Lindenthal was adamant, and by the end of March, Ammann had left his firm.[44]

During the summer and fall of 1923, Ammann made little progress in gaining support for his own bridge. The possibility of a vehicular tunnel at 110th Street, financed by investors, appeared to be of greater interest in the Bergen area, and a private association had begun to raise funds for that enterprise—which would almost certainly kill any prospects for a Fort Lee bridge, at least in the near term. Other investors began to look at 125th Street, and at 40th Street, as possible sites for private undertakings. But the tunnel investors, and their supporters in the New Jersey legislature, soon ran into double-barreled opposition: the governor of New York, Al Smith, said that he was strongly opposed to *any* private tunnels or bridges across the Hudson River; and the recently created Port of New York Authority objected, noting that it had developed a set of tunnel projects too, and arguing that all river crossings should be constructed as part of a comprehensive plan for transportation in the bistate region.[45]

■ **Helping the Port Authority
    to Redefine Its Goals**

In November 1923 the Port Authority announced that it would hold a public hearing on "the proposed additional vehicular tunnels." Now Governor Silzer and his bridge-building adviser saw an opportunity to enlist the Port Authority for service in their campaign. That agency, which had been created in 1921 primarily to help solve a freight railroad problem, had studied Lindenthal's railroad-bridge plan in 1921-1922 and rejected it as infeasible. As a result, the port commissioners and staff thought in terms of *tunnels*, which could be used to bring the rail lines from New Jersey and the West under the Hudson River and into Manhattan and Brooklyn. But bridges as well as tunnels were (at least in the abstract) in their domain; and though they thought mainly about railroads, they also had some interest in freight movement by truck— and trucks could travel on bridges as well as in tunnels.

On November 20, Silzer met with Julian Gregory, an influential member of the Port Authority board of commissioners, and said that he thought it unwise for the Authority's hearing to be limited to tunnels. That restriction, Silzer argued, "might be construed as limiting him [the governor], and the Port Authority, exclusively to tunnels," whereas he was "open-minded to any bridge proposition that might come forward." Indeed, the governor told Gregory, "he understood there was a strong sentiment on the part of some in favor of a bridge across the Hudson River." At the board meeting the next day, Gregory summarized Silzer's views and suggested that the December hearing be expanded to include the question of bridges across the Hudson as well as tunnels—and the Port Authority board agreed (Port of New York Authority, 1923).

The Port Authority's hearing was scheduled for December 5, and during the intervening weeks, Ammann worked furiously to strengthen the analytical case for a bridge at Fort Lee. His calculations indicated that a single-span bridge at that location would cost no more than $30 million—in contrast to more than $500 million for the Lindenthal project. Moreover, the immediate cost could be reduced to $25 million if the electric railway tracks he had included in the design were deferred until later.

Based on existing ferry traffic and studies done by the recently formed Committee on the Regional Plan, Ammann then estimated that 3 million vehicles would use the Fort Lee bridge in the first year—enough to meet all annual charges, if a reasonable toll charge were levied. Assuming continued increases ("in a few years the traffic should treble," he told Silzer), capital costs could also be paid off, and the bridge would in time be self-supporting.

He then suggested to Silzer that they talk with "some of the prominent bankers" to see if private investors might be willing to underwrite such a bridge, and Silzer sent Ammann to talk with Dwight Morrow of the J. P. Morgan firm. Meeting in early December, Morrow and Ammann agreed that the bridge might well be self-supporting, but Morrow doubted that adequate private capital could be attracted; both wrote to Silzer to recommend that *public* funds be used—either state moneys, or bonds floated by the Port of New York Authority.[46]

The Port Authority's public hearing strengthened Ammann's position. Most speakers agreed that more vehicular crossings of the Hudson were needed; and although there was support for new tunnels below 57th Street, the prestigious Committee on the Regional Plan and other speakers argued for a bridge farther north. The crucial question was *who* would take responsibility for such a bridge, and here Ammann's own views were clear. "The most practicable way" to proceed, he wrote to Silzer the day after the hearing, would be to have the Port Authority take on the challenge—which would also allow that agency (so far with no construction or operating projects at all)—to "test its working ability." Therefore, he urged Silzer to place Ammann's Fort Lee plan "at the earliest possible moment before the Port Authority."

A few days later the two men talked by telephone, and Ammann suggested that the Port Authority should be asked to make definitive studies not only of the Fort Lee plan but also of other interstate crossings that had been proposed—bridges from Perth Amboy and Bayonne to Staten Island, for example, and a scaled-down version of the 57th Street span. If the Port Authority were to conduct such studies, Ammann noted, it would need an expert bridge engineer, and "I shall be frank in stating that I should be glad to occupy such a position."[47]

Ammann and Silzer had agreed that Ammann would put together an extensive report on the Fort Lee project—covering technical engineering issues, traffic projections, financing questions, and probable impact of the bridge on regional development—and on December 17, Ammann's 22-page analysis reached Silzer's desk. That afternoon the governor forwarded the report to the Port Authority, with a letter from Ammann which concluded that the Fort Lee bridge could be paid for in 20 years. Silzer also released a public statement on his actions, noting that the Ammann plan was consistent with his own 1923 inaugural statement on the need for more Hudson crossings, and suggesting that the Port Authority could finance the Fort Lee bridge by issuing tax-exempt bonds—with "ample security" provided by tolls on the bridge.

The governor's efforts for the day had not yet ended. He also wrote a private letter to Commissioner Julian Gregory at the Port Authority, suggesting that, in carrying out its studies of the various bridge plans, the Authority might want to secure the services of "such a man as Mr. Ammann, who is thoroughly skilled in this kind of work."[48]

Gregory responded quickly, expressing his personal preference for bridging the Hudson at some point north of 125th Street. He also noted that Port Authority officials were now considering whether they should continue to focus their energies so strongly on moving freight by rail—or whether there might be a large role for trucks, in tunnels and over bridges. The Authority's staff then reviewed Ammann's report, and on December 21, the Commissioners reported to the two governors that they would carry out a detailed study of the Fort Lee plan.[49] Perhaps the Port Authority would now join Ammann in embracing the new automotive age.[50]

Ammann's efforts were beginning to bear fruit. But the events thus far brought a measure of pain as well as pleasure. Most hurtful was the behavior of Lindenthal. Silzer had sent him a personal copy of Ammann's detailed report on the Fort Lee plan, and on December 20, the great engineer responded with a letter condemning his former assistant, and alleging that Ammann had stolen Lindenthal's own ideas:

> Mr. A. had been my trusted assistant and friend for ten years, trained up in my office and acquainted with all my papers and methods. But I know his limitations. He never was necessary or indispensable to me. . . . Now

it appears that A. used his position of trust, the knowledge acquired in my service and the data and records in my office, to compete with me in plans for a bridge over the Hudson and to discredit my work on which I had employed him. He does not seem to see that his action is unethical and dishonorable. . . .'[51]

In addition, because the Port Authority now agreed to make a close study of Ammann's proposal, he waited for a call to join the Authority's staff—and take part in that study—but in vain. Christmas came and went, and Ammann was still an unemployed engineer.

■ On the Campaign Trail

Unemployed, but with much to do. The Port Authority would study his Fort Lee plan and would, he hoped, find that his engineering design and his analysis of costs, traffic flows, and financing were sound. But Ammann knew that the bistate agency—created to solve railroad problems, and staffed by railway engineers and statisticians—would be far more likely to take the next step and agree to build the great bridge if it found a groundswell of popular support for Ammann's 3,000-foot span. Moreover, legislative approval and probably some initial state funding would be required to get the project under way; here again, Trenton and Albany would be much more willing to commit their funds and the Port Authority's efforts to this project if local groups on both sides of the Hudson demanded action, for a bridge they felt was sorely needed.

So Ammann once again threw his energies into the effort to organize public support for the Fort Lee enterprise. Between late December of 1923 and April 1924, he held dozens of meetings with chambers of commerce and other groups in Bergen and nearby Passaic and Morris counties in New Jersey; he wrote to and visited similar associations in the Bronx, Harlem, Washington Heights, Westchester, and Yonkers in New York State; and he traveled into Connecticut and explained his arguments for the Fort Lee bridge before the Civil Engineering Society of that state.

By February 1924, he had developed working sketches of the proposed bridge—with its thin, graceful roadway, and its great towers, which would be vast metal structures sheathed in monumental stone. And now, when he spoke, Ammann could show his audiences some visual hint of his own deep motivations, which lay beyond engineering technique, beyond matters of practicality. It was true that the bridge would be a major engineering achievement; moreover, it would have a great impact on the efficiency of travel across a wide region, and so it would provide real benefits for residential choice, recreation, and economic growth. But a vast structure like this could also be—*should* also be—a work of art, and here was a large part of Ammann's incentive as he worked, without pay, to design and encourage the building of the Fort Lee bridge. Years before, reflecting on the Hell Gate Bridge, a monument to Lindenthal's own esthetic imagination and engineering skill, Ammann had argued that

> a great bridge in a great city, although primarily utilitarian in its purpose, should nevertheless be a work of art to which Science lends its aid. An elaborate stress sheet, worked out on a purely economic and scientific basis, does not make a great bridge. It is only with a broad sense for beauty and harmony, coupled with wide experience in the scientific field, that a monumental bridge can be created. (Ammann, 1918)

Now he might have the opportunity to create such a bridge, if the public and the state legislatures would approve it—and if some other engineer, of "greater reputation," were not chosen.

With increasing evidence of local support for the project, Ammann met with the state senator from Bergen County, William Mackay, and found him ready to press for legislation authorizing the Port Authority to construct a bridge at Fort Lee and smaller spans between Staten Island and New Jersey. During the spring of 1924, the New Jersey legislature took the first bite, endorsing Port Authority study and construction of two Staten Island bridges, and New York State approved similar legislation. In late May Governor Silzer sent a brief note to the Port Authority's general counsel, Julius Henry Cohen: "It has just occurred to me, in connection with the two bridges over Staten Island and your other bridge work, that the Port Authority ought to avail itself of the services of Mr. O. H. Ammann. . . . I understand that just at the moment he is available. . . . "[52] Just at the moment, and for more than a year now. But still the Port Authority did not call.

Although Ammann was willing to continue his organizing efforts, he had hoped that the various local groups might form a citizens' association to press for the bridge project. In April the Englewood Board of Trade and other business groups in the Fort Lee area seemed ready to create such an organization to coordinate efforts throughout northern New Jersey. Ammann then took his case across the river and was gratified when, early in May, the Harlem Board of Commerce—which had been a strong supporter of a tunnel at 125th Street—shifted its position and unanimously endorsed the Ammann bridge. Back in New Jersey, Ammann met with a committee of North Jersey mayors that had been formed to press for action on the bridge, and he drafted a strategic plan that the committee could use in gathering further support.[53]

Unfortunately, neither the business associations nor the mayors developed a viable organization, and neither was able to carry out a sustained campaign during the summer and fall of 1924. And Silzer told Ammann that he still did not think it desirable for him—a Democratic governor—to become actively involved in trying to organize support in Republican territory. So Ammann once again found himself taking the lead, engaged in a series of meetings with local groups and state legislators.[54]

By the fall the Port Authority began organizing to carry out the authorized bridge studies. During the autumn the agency sought bids for design work on the two Staten Island crossings, and Ammann responded. As he wrote to Governor Silzer on November 21: "I have submitted to the Port Authority a bid for the preparation of plans for the Arthur Kill Bridges and am now anxiously awaiting their decision." And while he waited, his political efforts produced important results: In late January 1925, the New Jersey Senate passed a bill authorizing the Port Authority to construct a bridge across the Hudson at Fort Lee, and the State Assembly soon followed suit. Ammann then crossed the Hudson to New York, where companion legislation had been introduced, and met with local and state officials, urging favorable action; and in late March, New York State approved the bill.[55]

■ From Political Entrepreneur to Bridge Builder

The Port Authority would now move forward—to build a bridge from Fort Lee to 179th Street, and to construct the two spans that had been authorized between New Jersey and Staten Island. But would Ammann have any role in their design and construction? In a letter on March 27, Ammann conveyed his concern to the governor. He expressed his hope that he would be asked to "take charge of the working out of the preliminary plans" for the Fort Lee bridge, but he thought there would

be opposition, and that an engineer "with long practice and wide reputation" might be selected instead. Reviewing his many activities on behalf
of the bridge project, Ammann concluded that he would appreciate
"anything you may be able to do to help" achieve a favorable outcome.

Two weeks later, with the Port Authority bills now signed in Trenton,
Governor Silzer once again wrote to Julian Gregory, who was now
chairman at the Port Authority. Noting that the Authority would soon be
proceeding with the Fort Lee bridge, Silzer suggested that "you take into
consideration for the doing of this work the name of O. H. Ammann."
Silzer continued:

> Mr. Ammann was one of the pioneers in this project, has spent two years
> of his time in advising the public of its advantages, has drawn freely upon
> his own ability as engineer, and in every way has probably done more than
> any other one man to bring this bridge into being.[56]

Silzer also sent a copy of this letter to Chief Engineer William Drinker
at the Port Authority; and he sent a copy to Ammann, with a note: "I
have it in the back of my mind somewhere that Mr. Drinker had an
impression that you were an able assistant, but that you had not had the
experience to independently undertake work of this kind." The governor
suggested Ammann talk with Drinker about this impression.

A few days later, Ammann met with Drinker. He thought it was an
"encouraging interview," though it contained disappointing news: Drinker
told him that the Port Authority had concluded that their first projects—
the two Staten Island Bridges—should be awarded to "an engineer of
long established reputation."[57] The job went to an independent engineering firm on an outside contract.

However, in its short and thus far uneventful life, the Port Authority
had already begun to develop a few traditions—and one of these was a
preference for hiring its own engineers and other experts as regular
members of the staff, rather than relying heavily on outside contractors.
By late April the chief engineer (himself a railroad man) had concluded
that the Authority ought to hire an engineer with bridge-building experience, and—Chairman Gregory wrote to Governor Silzer—Drinker had
recommended Othmar Ammann for the post.

The Commissioners soon concurred with Drinker's recommendation, and on July 3, Ammann sent a letter to Silzer noting that he had
assumed his duties as "bridge engineer on the Port Authority staff" on
July 1, and thanking the governor for his "goodwill and efforts on my

behalf." The long and active campaign ended on a restrained note, with the governor's final letter to an engineer who had at last landed a job and would now have to show that he had the capacity not only to fight for—but also to build—a great bridge.

July 15, 1925

My dear Mr. Ammann:

I have your letter of July 3d, and am, as you know, pleased at your appointment, because I am sure that you will be of much service to the two states.

Yours very truly,
s/George S. Silzer
Governor

Mr. O. H. Ammann
Boonton, N.J.

■ Ammann's Engineering Triumphs and the Dying of Political Light

> Gongs in white surplices, beshrouded wails,
> Far strum of fog horns . . . signals dispersed in veils.
> *Hart Crane, "Harbor Dawn" (1966)*

And there the story ends, and Ammann is lost from sight. Not, of course, the story of Othmar Ammann, engineer and artist. This tale of Ammann had yet 40 years to run and would be filled with activity and achievement. But the story of Ammann's political efforts, and their important role in permitting him to exercise his engineering genius— that story was at an end in 1925; and within a few years it was erased from the historical record.

The Triumphs

With the Port Authority now committed to his dream, the engineer poured his energies into the tasks before him—completing the detailed design for the Hudson River bridge, organizing the staff and consultants to carry out that large effort, and at the same time supervising the construction of the two smaller bridges authorized in 1924. In 1925 the legislatures authorized a third Staten Island span, from Bayonne, New Jersey, and so Ammann—while building the longest suspension bridge in the world at Fort Lee—also designed what would turn out to be the longest arch bridge in the world for the Bayonne crossing. The Fort Lee span, later renamed the George Washington Bridge, was initially expected to be finished in mid-1932; but the Port Authority's political insulation, and the willingness of the agency's senior executives to allow Ammann to set his own course, permitted him to exercise his considerable organizing talents once again. The effort moved ahead with unusual speed, and the crossing was completed 8 months ahead of schedule—and at a cost far below the 1925 estimate. The Bayonne Bridge was also opened in 1931, ahead of schedule and below cost; and both bridges won applause from engineers and observers for their engineering and esthetic merits.[58]

These efficient and spectacular achievements stamped the Port of New York Authority as a strikingly effective organization and, especially in view of the halting efforts of its rival—the joint state Bridge and Tunnel Commissions—ended the debate on how bridges and tunnels in the New York region should be constructed and operated.[59] With the support of the governors of both states, the Port Authority absorbed the joint commissions' staff in 1931, and its lucrative project, the Holland Tunnel; and the Authority was authorized to begin work on a second tunnel, to mid-Manhattan.[60] Had the port agency produced a stumbling or mediocre performance in these early bridge projects, it is quite possible that the Bridge and Tunnel Commissions would have retained control of the Holland crossing; and if that had occurred, the Port Authority would have faced a very uncertain future.[61]

What Ammann and his engineering team had done, therefore, was to ensure that the Port Authority would have the reputation for effective action and the strong financial base that would permit it, in the 1940s and 1950s, to reach out into new fields—into airport activities, marine terminals, and other urban development enterprises. And meanwhile, in the 1930s, Ammann would continue his engineering achievements, supervising construction of the Lincoln Tunnel, helping to design the Golden Gate Bridge, and, working with Robert Moses, creating a series of important bridges for his Triborough Authority.

History Rewritten

The story that ends in 1925 is the tale of Ammann the political entrepreneur, as his energies were absorbed in the technical and administrative activities summarized above. Because Ammann would hence-forth be associated with the Port Authority's leaders and with Robert Moses, with men and women who could take the political lead (and the political heat) while Ammann toiled in his favorite garden of structural art, that ending should not be surprising. What is remarkable, however, is that the story of how Ammann got there—into his favorite garden, as a dominant structural artist and engineer—was soon lost from sight; and the story of the Fort Lee bridge and the Port Authority's change in direction was rewritten without him, or nearly so.

The fading of his political light did not occur at once. In *The New Yorker* profile in 1934 (MacKaye, 1934), Ammann's role in the development of the great bridge is briefly described: "He prepared his blueprints and became his own advocate; he spoke before public gatherings and interviewed public officials. Eventually he convinced Governor Silzer of New Jersey that the two adjoining states should erect the bridge" (p. 25). However, in that essay and in other early articles, Ammann's personality and his reputation are described in ways that must soon overshadow all else—making it difficult to imagine Ammann appealing for support among local gatherings in Teaneck and Leonia, or making his way by ferry, sketches of his bridge under his arm, to dark halls in Washington Heights and the Bronx, or draining his savings while waiting for others to act. By 1934, Ammann is "one of the immortals of bridge engineering and design, a genius." What kind of a man is he? He is "quiet . . . tactful and courteous in his relationships with other men, calm in his judgments, flawless in his engineering." In "contrast to the mightiness of his work," Ammann "is quiet, mild spoken and retiring."[62]

By the 1940s, Ammann's entrepreneurial role in the origin of the George Washington Bridge vanishes from sight. In *Bridges and Their Builders*, David Steinman and Sara Ruth Watson (1941)—whose goal is to tell the story of great bridges as "an epic of human vision and courage, high hopes and disappointments, heroic efforts and inspiring achievements"—offer rich descriptions of the campaigns to obtain approval of Brooklyn Bridge, the Golden Gate, and many others. But of the Fort Lee bridge they tell the reader only that Silzer advocated it in 1923, that he and Al Smith wanted the Port Authority to build it, and that "two years later" the state legislatures gave their approval.[63]

A far more detailed description of the political activities leading to approval of the George Washington span is provided in Jacob Binder's 1942 book. Binder was an active member of the Bergen County coali-tion that came to life in 1924-1925 to press for state approval, and his lengthy discussion conveys a sense of authenticity. In his account, however, Ammann's contribution is that of dedicated bridge designer; others do all the political work.[64]

In the 1940s, two other books treated Othmar Ammann and the George Washington Bridge in some detail. Both books were concerned with the political forces that shaped the creation of the Port of New York Authority and its first decades of operation, and Ammann was once again relegated to his engineering role. In his classic volume, *The Port of New York Authority*, Erwin Bard (1942) described the efforts of Al Smith to persuade the Port Authority to take on vehicular projects, and (as noted earlier in this chapter) Bard indicated that the Authority's staff was at first not eager to add that task to its rail-improvement plans. Then, in 1924-1925, the state legislatures authorized Port Authority action to construct three Staten Island bridges and a Hudson River span. At that point, negotiations with the railroads were collapsing, and within the Port agency, Bard noted, "the center of gravity began shifting to vehicular traffic," because bridge building "offered a chance" to show that the agency could accomplish something.

So the Authority's staff was "reshaped to fit the needs of construction," and "Othmar H. Ammann was engaged as Bridge Engineer." Of Ammann's background, Bard says only this: "Coming to the Port Authority with no great reputation, he became widely known as designer of its bridges and head of its Engineering Department."[65]

The other volume was written by Julius Henry Cohen (1946), author of the legislation creating the Port Authority as well as its general counsel from 1921 until his retirement in 1942. During his years with the port agency, Cohen kept careful track of the political activities affecting the Port Authority.[66] Therefore, it would seem unlikely that he was entirely ignorant of Ammann's crucial political role in the campaign for the Fort Lee bridge. Yet Cohen's discussion of Ammann omits that effort, and in fact Cohen cites Ammann to illustrate the gulf that separates the methods of the engineer from those of the political leader.[67]

Since Ammann's death in 1965, several extensive biographical essays on his life and work have appeared. Almost all have been written by engineers who, perhaps naturally, have concentrated on Ammann as bridge designer and engineering administrator. They do offer brief hints of the "crucial years" of 1923-1925, but little more. In a 1974 biography of Ammann, for example, Fritz Stussi reports that Ammann left Lindenthal's employ in the spring of 1923, that he submitted a detailed report on the Fort Lee bridge to Silzer in December 1923, and that he was hired by the Port Authority sometime in 1925 (Stüssi, 1974, pp. 13-14, 16, 46).

Urs Widmer's perceptive 1979 essay provides detailed information on Ammann's early training and on his engineering activities. He also notes that the break with Lindenthal occurred in 1923, in part because Ammann feared that the campaign for tunnels would soon sweep aside the possibility of a bridge across the Hudson.[68] And Widmer's article provides one of the two best summary descriptions of Ammann's activities in the 2 years after he left Lindenthal.[69] The other is found in a brief essay by Ammann's daughter.[70]

In the most recent paper on Ammann's work, by Edward Cohen, the missing years are again touched on briefly. Cohen indicates that Ammann left Lindenthal after a spring 1923 argument about trimming the size of the Hudson bridge, and that Ammann then worked on his own for 2 years, designing a more modest span, which would be attractive to Governor Silzer and others in terms of financial cost and esthetic appearance. In 1925, with the Port Authority authorized to construct the bridge, it thereupon "appointed Ammann Bridge Engineer."[71]

■ Why the Political Entrepreneur Was Lost

So Ammann's early role as political organizer has continued in eclipse. There are undoubtedly several reasons for this gap in the biographical and historical record, some of them specific to individual authors.[72] I am inclined to place considerable emphasis, however, on Ammann's position as exemplar of the engineering profession in its ideal form. That image of Ammann has, I would argue tentatively, tended to drive out any close consideration of other important aspects of Ammann's talents and behavior, even by those—including Widmer and Cohen—who have had access to a fair portion of the evidence regarding Ammann's political activities in the early 1920s.

What I mean is this: When one reads the engineering literature, one learns that the highest standards of the profession (perhaps particularly of the profession of civil engineering) are *efficiency*, *economy*, and *grace*, captured in a structure that is actually *built*, and that works.[73] The goal of efficiency entails "a desire for minimum materials, which results in less weight, less cost, and less visual mass." The discipline of economy means a desire for simplicity in construction as well as "a final integrated form." The search for grace, or engineering elegance, involves the visual expression of efficiency and economy "through thinness and integration," and through contrast with the surrounding environment.[74]

All of Ammann's engineering achievements—from his designs for the George Washington and Bayonne bridges in the 1920s through his Verrazano-Narrows span in the 1950s—emphasize these values at a very high level of distinction. (For an extensive discussion of these aspect of Ammann's work, see Billington, 1983, pp. 130-134, 137-140.) Moreover, his writing—which also strikes a high level in clarity and detail—underscores the importance of these values in his own work.[75]

But Ammann was exemplar for a profession for reasons that go beyond the "daring elegance" of his designs, and beyond the fact that these structures were built at low cost and remain standing. To those who knew him, he also personified the traits of character that had stamped the best members of the engineering craft extending back into the nineteenth century and beyond. As John Jervis, builder of the Erie Canal and other large projects of that era, had written:

> A true engineer, first of all, considers his duties as a trust and directs his whole energies to discharge of the trust. . . . He is so immersed in his profession that he has no occasion to seek other sources of amusement, and is therefore always at his post. He has no ambition to be rich, and therefore eschews all commissions that blind the eyes and impair fidelity to his trust.

And, like the engineers of the earlier period, Ammann was "independent, austere" and "self-confident" (Morison, 1974, pp. 68, 93).

In addition, Ammann's distinctive abilities and personality were underscored by contrast with some other prominent bridge builders of the twentieth century—Lindenthal and Steinman, whose bridge-tower embellishments did not perhaps reach the high standards of economy and grace found in Ammann's structures, and especially Joseph B. Strauss, the chief engineer for the Golden Gate Bridge. Any story of the building of that great structure would have to devote considerable attention to Strauss. But because Strauss was widely understood to be a bridge designer of very modest capacities, any effort to examine why he headed the engineering team, and how the bridge was designed and public support obtained, would soon lead the historian into the complex story of Strauss as political entrepreneur—the field in which he made his major contributions. (For a detailed discussion of Strauss and "his" bridge, see van der Zee, 1986.)

In contrast, from the early 1930s onward Ammann was viewed by members of his profession and by the wider public as "one of the immortals of bridge engineering and design"; and his quiet manner and self-assurance reinforced the perception that this was an engineer's engineer. Therefore, when the historian or engineer asks the question, "Why was Ammann chosen to design the George Washington Bridge?", the answer may seem self-evident: The Port Authority chose Ammann because he was the best man for the job. And so, in exploring why the Port Authority shifted gears to take on this task, and how public support for building the bridge was obtained, the researcher easily passes by Ammann, and looks for the answers in the activities of real estate developers and other interest groups in the region, and in the imperialistic visions of an ambitious Port Authority (see, e.g., Danielson & Doig, 1982, pp. 186-194). Not that these factors are irrelevant, but it now seems clear, I think, that they do not provide an adequate explanation.

■ **The Study of Political Power
and the Role of Biography**

In their efforts to understand how political power is organized and used in a society, political scientists have generally directed their energies toward examining the actions of interest groups and the behavior of government bureaucracies. With few exceptions, that perspective treats the role of individuals as insignificant in shaping government policy and the use of public resources—except insofar as individuals act "in role," as members and leaders of pressure groups and bureaucracies whose aim is to maximize the economic profit or other goals of their own, narrow organizations.[76]

The general argument illustrated by this study of the "George Washington Bridge case" is that the traditional political science perspective is too narrow, and that close biographical studies—scrutinizing the evolving perceptions, motivations, and activities of specific individuals —will often be rewarding, opening up lines of inquiry that extend beyond role-bounded behavior and the kind of reductionism that the traditional mode of inquiry often entails. In this case, the evidence drawn from a biographical study of Ammann—joined with an exploration of institutional and other factors—indicates that Ammann's activities in the 1920s were influential in several directions. Some of these are noted in earlier sections of the chapter; let me at this point identify these areas of influence and suggest their relationships with broader forces at work.

Ammann's Impact on Urban Development
in the New York Region

If one asks, for example, why the vast reaches of Bergen County and nearby areas in northeastern New Jersey remained a rural enclave until the 1920s, while areas on Long Island equally distant from Manhattan had become densely populated, it seems clear that the absence of bridges and tunnels connecting North Jersey to Manhattan's major employment centers was a major factor. Geography and technology joined forces here, because the Hudson River was far wider than the waterway separating Long Island from Manhattan. East of Manhattan, technological advances had permitted bridges to be cast across the East River beginning in the 1880s, supplanting the several ferry routes; and as a result, the pressure for suburban living soon sent the population flowing into Brooklyn, Queens, and Nassau County. As the automobile grew in popularity after 1910, residences then spread widely across Long Island, far from the rail lines that crossed the river.

By 1910, however, geography no longer stood in the way of bridges and tunnels across the wide Hudson. Engineering advances would permit bridge spans of 3,500 feet and longer; and—except for some uncertainty about how to remove carbon monoxide—engineering techniques would also permit long vehicular tunnels under rivers like the Hudson. Widespread popular demand was present, too; automobiles and trucks crossing the Hudson by ferry sometimes had to wait several hours in long lines at ferry terminals.[77]

What was missing in 1910 and subsequent years was the organizational and political capacity to span the Hudson. Two states had to agree on where to locate interstate bridges and tunnels, on private versus public financing, and on what governmental bodies would build or monitor the building of these major arteries of commerce and communication. By 1919, one vehicular tunnel had been started, haltingly, by two state commissions working uneasily together; but that tunnel would provide only two lanes in each direction, and projections indicated that three or four times that capacity was needed. Should there be two or three more tunnels, distributing traffic (and suburban population growth) across Union, Essex, Morris, and southern Bergen counties? Or should there be one great bridge with capacity equal to all those tunnels—and if a bridge, should it lead into midtown Manhattan, or be farther north, shifting population growth to northern Bergen and Passaic counties, and perhaps diverting some traffic away from congested mid-Manhattan?

Local government officials and real estate developers at the terminus of a tunnel in Jersey City or Weehawken could see the advantage in tax ratables if one of the river crossings was located near that local community, so an incremental series of tunnels was probably likely to be the result of interest group and local community pressures. It was more difficult to gather the political support needed for a large bridge; a bridge would cost more than any single tunnel, and that meant that political (and perhaps financial) support would have to be gathered more widely—which was difficult especially in Bergen County, which had dozens of small towns and little tradition of cooperation among its towns and villages. And to build a bridge with 8 to 12 lanes meant that there might be no need for tunnels for a long time; so the bridge was a threat to those who wanted tunnel crossings near their own communities.

To simplify only slightly, what Ammann did was to alter the political environment significantly, perhaps dramatically. He wanted a great bridge flung out from the high Palisades. He had wanted one since 1904; it was an engineer's dream. When Lindenthal proved unequal to the task of understanding the complex reality of the political environment, Ammann broke free. By the start of 1923, he had achieved the first step—convincing one state governor that a vehicular bridge at the Palisades was feasible and desirable. During the next 2 years he overcame the political fragmentation and mutual suspicion that made cooperation among local civic groups and political officials so difficult. He had an idea, he could show it visibly and dramatically, and he was persuasive in arguing that the citizens of Englewood and Teaneck and Boonton—and their counterparts in the North Bronx and Westchester—should exercise their political muscle in order to help accomplish his dream, rather than waiting in long ferry lines until someone else could put another tunnel down, under the mud.[78]

Perhaps most important—in terms of his short-term and long-term impact on the New York region—Ammann exercised a profound influence on the direction and the reputation of the Port of New York Authority. He did this in part by recognizing that the Port Authority—in contrast to the joint Bridge and Tunnel Commissions—had the political characteristics required to get the great bridge built, and to get it built efficiently.[79] And if the Port Authority had not been created precisely to build large vehicular bridges, it still might be persuaded to do the job—especially if his ally, the governor, could help to stretch the Authority's collective mind. In the long run, of course, Ammann's influence on the port agency depended on his performance as a bridge builder and administrator; the argument regarding that role and his impact are set forth in the previous section. There are other factors, certainly, that come into play in mapping the causes of the Port Authority's expanding domain and power in the 1940s and beyond. However, Ammann's early efforts, through 1931, probably rank as *necessary* if not sufficient in understanding why that agency would later prosper—why it had the reputation and funds needed to permit it to take over the region's airports and build the world's largest bus terminal in the 1940s, and then, in the 1950s, to define and take the lead in meeting the region's highway needs, defeating Robert Moses when he resisted, and joining with him when that strategy was consistent with the Port Authority's vision of how to shape the bistate region. (For a review of the Port Authority's evolution after 1931, and of the conflicts with Moses, see Doig, 1990, pp. 205ff.)

Benefits and Dangers

What is true in this case applies to other cases as well. That is, biographical analysis often adds an important dimension of understanding regarding the uses of political power and the evolution of social policy. We have biographical studies now, for example, that advance our knowledge of how the American navy came to accept—though reluctantly—major improvements in gunfire technology and technique in the early 1900s; of why the American Social Security system evolved as it did from the 1930s to the 1980s; and of the factors that were crucial in the creation and evolution of the U.S. Forest Service.[80] We could use more such studies, particularly in fields in which interest groups, bureaucratic incentives, and underlying economic and social forces are generally viewed as providing an adequate understanding of the patterns of power and the structure of social outcomes.

A further point worth noting here is that a sustained effort of biographical analysis tends to stretch the mind of the researcher, suggesting additional perspectives from which a problem can be and perhaps ought to be studied. The biographer needs to be able to see the situation from behind the eyes of the (political) actor, and that effort leads one to attempt to understand the situation as it was perceived by those involved—at a time when the future was unknown and obstacles that now seem unimportant loomed large. As a consequence, the researcher may uncover important causal variables that are not readily identified when viewing the problem from the outside, when the future is known. Interviews and written records not motivated by biographical analysis can be used in this way too, of course, but the biographer may be more likely to immerse herself or himself in the situation as it was seen and felt by participants at the time, and therefore may be able to unearth additional lines of causal inquiry.[81]

In addition, biographical studies may be helpful in encouraging individuals to find meaning in their lives—by suggesting that individuals can act, even if they live in complex and fragmented societies, with some hope of making a difference. As Jean Strouse (1988), biographer of Alice James, notes:

> Future historians may characterize the late twentieth century by its sense of fragmentation, its lack of confidence in history's progress, its loss of consensus about what an "exemplary" life might be. People still long for models of wholeness, though—for evidence that individual lives and choices matter.

Biographical studies, she argues, may provide useful cases to illustrate those positive themes (Strouse, 1988, pp. 184-185).

Having been absorbed now for several years in a set of three intertwined biographical studies, I should conclude on a cautionary note. The biographer needs to be alert to the danger that the subject of his or her attention may threaten to swallow the researcher. Lippmann's biographer recalls his concern partway through his many years of work: "I came to fear the way in which he would insidiously take over my life—take it over in time, until I often felt I hardly had any life outside of Walter Lippmann, and also by forcing me constantly to define myself in terms of him and him in terms of me" (Steel, 1988, p. 124). Equally important, all biographers run the risk of becoming so attracted to the subject, or so repulsed, that the objectivity essential to careful analysis is lost.[82] As I argue elsewhere, this is a danger well illustrated by Robert Caro's important but one-sided biography of Robert Moses (see Caro, 1974; Doig, 1990, pp. 225-227). It is a danger not easily avoided, but perhaps less likely for those, such as political scientists, who use biographical analysis as an adjunct in probing patterns of power and influence, than for the researcher whose central goal is biography.

NOTES

1. On their impact on residential and employment patterns in the New York region, see Danielson and Doig (1982, chap. 6).
2. See "A Monumental Bridge" (1927) and Wisehart (1928).

3. As David Billington (1983, p. 129) comments in his recent volume: "No twentieth-century engineer has left more of a mark on steel bridge design than Othmar Ammann. Taken as a whole, his designs . . . provide the best example of structural steel bridge art done in this century."

When the American Society of Civil Engineering prepared to commemorate its 100th anniversary in 1952, it chose an Ammann creation; the U.S. stamp that year honoring the Society shows a wooden covered bridge in the lower left-hand corner, with the George Washington span extending across the main body of the stamp.

4. See in particular the items listed below in notes 8, 10, 14, and 19.

5. The entrepreneur identifies new goals, works out the steps essential to achieving these goals, and marshals the resources needed to move forward toward his or her preferred ends. Generally the entrepreneur's efforts require overcoming hostile forces and other obstacles, and they often entail some risk to the entrepreneur's financial security and career. See Doig and Hargrove (1987, pp. 7-8) and sources cited there. On the concept of the "political entrepreneur," see also Mollenkopf (1983, p. 6ff).

6. The professor, K. E. Hilgard, had worked as a railroad-bridge engineer in the United States, and he pressed Ammann to go to the United States, where "the engineer has greater freedom in applying individual ideas" and where young men were sometimes put in charge of work "which, in Europe, only graybeards would be allowed to perform" (quoted in Wisehart, 1928, p. 183).

7. Ammann later reported that "my first serious interest in the problem of bridging the Hudson was awakened shortly after my arrival in New York," when he visited the top of the Palisades cliffs on the Jersey shore across from Manhattan. "For the first time I could envisage the bold undertaking, the spanning of the broad waterway with a single leap of 3000 feet from shore to shore, nearly twice the longest span in existence. . . . From that moment . . . I followed all developments with respect to the bridging of the Hudson River with keenest interest" (quoted in Widmer, 1979b, pp. 5-6).

During these years, the only way that horses and motor vehicles could travel between New Jersey and New York City was via ferry—unless they journeyed 50 miles north, where the Hudson was much narrower and smaller bridges had been constructed. Travelers without horse, auto, or truck could cross the Hudson as railroad passengers, once the Pennsylvania Railroad tunnel and two smaller rail tunnels under the Hudson were completed in 1908-1910. The question of whether and how a bridge might be cast over the Hudson had been debated sporadically since the early 1800s.

8. The Hell Gate crossing, to span the East River between Queens and the Bronx, would fill the major gap in the Eastern rail system, allowing railroad trains to travel from New England to the network of tracks that connected New York to New Jersey and the continental rail systems. The Hell Gate would be the longest arch bridge in the world (see Billington, 1983, pp. 125-128).

9. The proposed bridge, as Lindenthal sketched it out in 1920, included 12 railroad tracks and 20 vehicular lanes on two levels, carried by a single gigantic span across the Hudson. The price tag was $180 million or more (see Widmer, 1979b, p. 10; Shanor, 1988, p. 142).

10. On Ammann's career at the Port Authority and subsequently, see Katz, L. (1979, pp. 20-21; 1988, pp. 33-39), Stüssi (1974, p. 46ff), Widmer (1979a, pp. 38-94), and the papers by Widmer and Cohen cited earlier.

11. On the highway studies in the 1950s and the earlier battles, see Doig (1990, pp. 209-225).

12. Ammann wrote on November 21, 1924, to George S. Silzer, "I have submitted to the Port Authority a bid for the preparation of plans for the Arthur Kill Bridges and am now anxiously awaiting their decision." When he met with the Port Authority's chief engineer on April 17, 1925, however, he learned that because these would be the Authority's "first work, it appeared advisable to give it to an engineer of long established reputation" (Ammann to Silzer, April 17, 1925). The two bridges over the Arthur Kill—a narrow waterway between New Jersey and Staten Island—are the Goethals and the Outerbridge Crossing; they were designed by two private consultants whose names are now largely erased from the Port Authority's institutional memory. Ammann is often listed incorrectly as the designer of these two metal monsters.

13. Trucks and passenger automobiles could cross the Hudson only by waiting in long lines for a ferry, or by traveling dozens of miles north to pass over the narrower Hudson via bridge near Poughkeepsie.

14. For summaries of these developments, see Bard (1942, pp. 180-181) and Binder (1942, pp. 174-180).

15. Smith's favorable inclination toward the Port Authority was also shaped by his earlier involvement: During his first term as governor (1919-1921), Smith had pressed for the creation of the bistate agency. Defeated for reelection, he had then been appointed by the new governor (Nathan Miller) in 1921 as one of the Port Authority's first set of commissioners. Elected governor once again, he took office in 1923, convinced that the Port Authority had an important role to play in overcoming the fabled inefficiencies in the New York region's transportation system (see Bard, 1942, pp. 32-33, 181-82; Doig, 1988, pp. 58-62, 72-76, 81).

16. On the Port Authority's attitude, and Silzer's views, see Bard (1942, pp. 182-185) and several items in the Silzer files in the New Jersey Archives: E. Outerbridge to G. Silzer, March 9, 1923; Silzer veto message on New Jersey bill for private bridges and tunnels, October, 1923; Ammann to Silzer, November 22, 1923; Silzer to D. Morrow, November 27, 1923; Morrow to Silzer, December 5, 1923; J. Gregory to Silzer, December 18, 1923.

17. The classic study of the Authority's first decades is Erwin Bard's 1942 book, *The Port of New York Authority.* Bard indicates that Governor Alfred E. Smith urged the Port Authority to take on bridge and tunnel projects for motor vehicles in 1923-1925, and that the Port Authority staff was reluctant to embrace that new task. But the possibility that the Authority could carry out its preferred program—large rail projects—faded by 1925-1926, because the railroads were unwilling to cooperate. A bridge-building program "offered a chance" for some kind of achievement in an organization that was floundering. At that point, Ammann "was engaged as Bridge Engineer." He arrived, Bard concludes, "with no great reputation" (pp. 185, 193). The book offers no information on why he was chosen.

Two other books—Cohen (1946) and Binder (1942)—discuss the political activities surrounding the Port Authority's Hudson bridge project. Both refer to Ammann's engineering activities at the Port Authority, but provide little information on why he was selected to take charge of the bridge project. Cohen (p. 123) appears to assess Ammann as a "pure engineer" who lacked the talents and inclination needed to be an adept political organizer.

18. The relevant materials from the 1920s appear to have been destroyed when the Port Authority offices were moved to the World Trade Center in the early 1970s.

19. In discussion, Billington agreed that it was surprising the Port Authority had chosen Ammann to carry out the George Washington Bridge project, rather than a more prominent engineer or consulting firm. Billington had the impression Ammann might have put his own name forward in some way, but he had no details. Billington has also written on Ammann; his book, *The Tower and the Bridge* (1983), discusses Ammann's engineering and his esthetics, and earlier, in 1977, he had written an article, "History and Esthetics in Suspension Bridges," which criticized Ammann's George Washington Bridge design on both technical and esthetic grounds, raising a modest flurry among the faithful.

20. A brief vignette on the search: When I called the State Archives in July 1987, I knew that the papers of many of the state's governors had not yet been organized, and I thought that Silzer's—like the gubernatorial papers of Woodrow Wilson and others— might still be in folders without subject headings. "You're in luck," a staff member responded when I called, "we've had a student here this summer, and he's just finished organizing Silzer's papers." I asked whether there were any files titled Hudson River Bridge, or George Washington Bridge. The answer was "no." Perhaps this was a dead end. Did she see *any* files on bridges? "Well, yes," she responded; "there's a very bulky file here labeled 'Ammann Bridge.' " I soon headed for Trenton.

21. The main span of a bridge at 179th Street would be 3,500 feet; the longest spans then in existence or under construction were the Manhattan Bridge (1,470 feet) and Williamsburg (1,600 feet), both across the East River to Manhattan, Bear Mountain Bridge over the narrower Hudson farther north (1,630 feet), the Delaware River Bridge to Philadelphia (1,750 feet), and the Ambassador Bridge over the Detroit River to Canada (1,850 feet).

The engineering advances that would, in Ammann's opinion, make possible this giant step in span length included the creation of new alloy steels, development of more accurate methods of shop fabrication and shop assembling of bridge parts, better methods of calculating stresses and of model experimentation, and an improved conception of how to evaluate the forces that stabilize (or "stiffen") a massive bridge. Ammann's views on these issues are set forth in several speeches and papers during the 1920s and 1930s; for a summary discussion, see his paper, "Brobdingnagian Bridges" (1931).

22. Silzer did not set down these views systematically in one place. However, I believe this is a fair summary of his thinking in 1922-1925. My main sources for his views are Silzer's inaugural address in January 1923, his public addresses in January 1924 and 1925, and the correspondence and newspaper clippings found in the "Ammann Bridge" file in the New Jersey Archives—particularly his letters of June 7, 1923 (to the managing editor of *The New York Times*); and November 27, 1923 (to Dwight Morrow); Silzer's public statement of December 17 on Ammann's plans; Ammann's letters to Silzer on December 12, 13, 17, and 24, 1923 (which summarize their several discussions); Silzer's letter to Port of New York Authority commissioner Julian Gregory, December 31, 1923; and a number of similar materials in 1924 and early 1925. Some specific examples will be given later in this chapter.

23. "Our neighbor knew how much midnight oil father was burning because she often had to attend to her sick mother during the night. 'Whenever I looked over to the Ammann house, at one o'clock, three o'clock, there was always a light burning in Mr. Ammann's study and I knew he was working.' " (Durrer, 1979, p. 29).

24. The quotation is from a 1907 paper by Burnham, a Chicago architect and planner, and is widely reprinted; it is quoted here from the frontispiece of a recent book (Shanor,

1988) that describes the wondrous hopes of Lindenthal and a large band of engineers and others who sought to reshape New York.

25. For this summary of Lindenthal's life and work, I draw mainly on Billington (1983, pp. 123-132) and Shanor (1988, pp. 136-149).

26. See Reier, S. (1977, pp. 41-57). The City engineers were under constant pressure to allocate contracts to firms associated with influential politicians. On the strategies used by Tammany Hall in obtaining Queensboro Bridge contracts in 1903, for example, see the summary of court hearings reported in "Dummies in City Contract" (1911).

27. The Hell Gate Bridge spans the upper part of the East River and is a key link between the rail lines in New England and the railroad system that extends from Manhattan under the Hudson River to the rest of the nation.

28. Lindenthal's aim was to finance the bridge project with private capital. However, he would need some governmental assistance—approval of the federal government for a bridge across a navigable waterway; and perhaps monopoly rights, in order to ensure investors in his bridge that toll revenues would not be drained off by any competing bridge to Manhattan.

29. Ammann to his mother, April 24, 1921 (trans. Margot Ammann, 1988).

30. Lindenthal was fairly explicit in sketching out his perceptions and values on these several issues, particularly in his reports in 1921-1922, and in his letters to Governor Silzer in 1924 and 1925. See specific citations later in this chapter.

31. Indeed one of those tunnels had been started already, in 1920, between Canal Street in lower Manhattan and Jersey City; and there were plans afoot to follow that effort with a series of other underwater crossings—which might in time (so argued tunnel engineers and their associates) entirely eliminate the need for bridges over the wide Hudson. Perhaps a bridge could be more esthetic—even a work of art; the tunnelers would admit this, though a look at the Queensboro Bridge suggested to some that such promise could easily be despoiled. Tunnels, however, were probably cheaper; and because they could be constructed at many locations up and down the River, they would disperse traffic, not concentrate it into one monstrous traffic jam, as Lindenthal's great bridge seemed likely to do.

32. Othmar Ammann to Rosa Labhardt Ammann, December 14, 1923 (trans. Margot Ammann, 1988).

33. On Silzer's values and programs, see his inaugural address (January 1923) and his first and second annual messages to the legislature (January 1924 and 1925); Kull (1930, pp. 1080-1083); and Stellhorn and Birkner (1982, pp. 194-196).

34. Ammann's diary entries during the years 1917-1920, when he was manager of the company, list several meetings with Silzer and other directors, including Lindenthal.

35. The tone of Lindenthal's January 30, 1923, letter to Silzer suggests that earlier discussions of engineering and financial details had taken place between the two men.

36. And so Ammann's 1904 vision might be reclaimed, and converted into steel. ("I could envisage the bold undertaking, the spanning of the broad waterway with a single leap of 3000 feet from shore to shore, nearly twice the longest span in existence"—Ammann's recollection of his visit to the top of the Jersey Palisades, shortly after his arrival in America; see Note 7, above.)

37. The effort to build a vehicular tunnel between Jersey City and Canal Street, started in 1919, had been carried forward with much quarreling and many delays by two state commissions, and in 1923 completion of the [Holland] tunnel was still years off. There

was also some concern that the ventilation system in the underwater tunnel would not carry off the carbon monoxide: Why build a second death trap, skeptics asked, until the first has been tried out?

38. The Port Authority had been created by the two states in April 1921 and was mandated to devise a plan to improve freight transportation in the bistate region around New York Bay. The agency's first plan, published in 1922, focused on ways to improve railroad service, including the construction of underwater rail tunnels between North Jersey and New York City. But the agency also viewed trucks as relevant to its task, serving as feeders between rail terminals and customers. So it might view a bridge at 179th Street as valuable for moving feeder trucks across the region.

39. The Swiss heritage included his student years in Zurich with Wilhelm Ritter, who emphasized esthetic as well as technical principles in bridge building. See Billington (1979, pp. 1103ff). And, at a genetic level, Ammann may also have drawn upon his maternal grandfather, Emanuel Labhardt, a well-known landscape artist. M. Ammann (c. 1979, p. 1).

40. Since early in the nineteenth century, engineers had attempted to make suspension bridges "more and more rigid, in order to eliminate the wavelike motion due to flexibility." Rigidity was obtained by using heavy trusses, which required large amounts of expensive steel. Ammann's studies convinced him that "for a long-span suspension bridge a rigid system was not necessary." By eliminating large, stiff trusses, Ammann reduced the cost of the bridge by about 15%. (The quotations are from the form nominating the George Washington Bridge to be a National Historic Engineering Landmark, as quoted in Katz, 1988, p. 34.)

41. As he wrote to his mother in 1921, "Toward strangers one is always covered with a veil. . . . The human soul must not expose itself to the profanities of the world" (February 12, 1921; trans. Margot Ammann, 1988). My comments on Ammann's personality also benefited from discussions with Margot Ammann and Sylva Brunner.

42. O. H. Ammann, letter to Governor George S. Silzer, January 9, 1923; from the George Silzer files in the New Jersey Archives, Trenton. In general, when the writer, addressee, and date of a letter are provided in the text, footnote references will be omitted below. The January 9 letter is the first correspondence between Ammann and Silzer in the Silzer files.

43. Silzer, Inaugural Address, January 16, 1923, p. 8. As his reference to "railroad terminal service" suggests, neither Silzer nor Ammann was yet ready to break entirely with Lindenthal; part of Lindenthal's plan was a railroad terminal at the Manhattan end of the 57th Street bridge.

44. Ammann reported one of their final exchanges in his diary entry of March 22, 1923: "Submitted memo to G.L., urging reduction of H.R. Br. program dated Mar. 21. G.L. rebuked me severely for my 'timidity' and 'shortsightedness' in not looking far enough ahead. He stated that he was looking ahead for a 1000 years" (quoted in Widmer, 1979a, p. 8).

"Lindenthal took the first opportunity to lay Ammann off," Edward Cohen concludes, "and in 1923 the two men parted" (Cohen et al., 1987, p. 9).

45. The Port Authority's 1922 Comprehensive Plan, which had been endorsed by the legislatures of both states, included an array of rail tunnels under the Hudson River and other waterways in the region. Governor Smith's opposition was motivated in part by his belief that the Port Authority should have control over all interstate tunnels and bridges, and in part by his general opposition to monopoly control by a private corporation over a

crucial transportation facility. Because New York State's official approval would be needed for a private tunnel into that state, the threat of a gubernatorial veto was an important obstacle for the private association. For further information, see Doig, J. (1988, pp. 17-21).

46. This summary is drawn from Ammann letters to Silzer, November 22, December 6, 1923; Silzer letter to Morrow, November 27, 1923; Morrow letter to Silzer, December 5, 1923.

47. Ammann to Silzer, December 12, December 13, 1923. (Having left Lindenthal's employ early in the spring of 1923, Ammann had continued to work full-time on the Fort Lee bridge project without pay, using his savings to support his family, during the remainder of 1923.)

48. Ammann to Silzer, December 17, 1923 (with attachments); Silzer to the Port Authority Commissioners, December 17, 1923; Silzer to Gregory, December 17, 1923.

49. Gregory to Silzer, December 19, 1923; Port Authority Commissioners, Minutes, December 19, 1923; Port Authority Commissioners, letters to Governors Silzer and Smith, December 21, 1923.

50. Both Ammann and the Port Authority had their roots in the era of railroads and rail freight, but Ammann had found it easy to respond—in his general thinking about transportation patterns, and in developing detailed designs—to the increasing use of trucks and automobiles. In part, this reflected his broad disposition to let his mind absorb new facts and use them to modify his views about the world (rather than reinterpreting new facts so they were consistent with his fixed views).

In addition, to a bridge engineer, designing wide spans for cars and trucks offered great advantages over railway bridges—for the structures could be lighter and less costly, and their location was not limited to the endpoints of existing rail lines. And to Ammann in particular, with his driving esthetic interest in constructing bridges that had a "light and graceful appearance," the automotive age offered possibilities for artistic achievement denied to those who built in the railroad era. (On Ammann's esthetic perspective, see Billington, 1983, pp. 128-146.)

51. Gustav Lindenthal to George Silzer, December 20, 1923. I have no direct evidence that Governor Silzer showed this letter to Ammann, but Silzer's general habit was to send letters he received to other interested parties, and Ammann's letters to Silzer in subsequent months make it clear that he knew Lindenthal had criticized his behavior on professional and personal grounds. As earlier sections of this chapter indicate, Lindenthal's characterization of Ammann's actions was quite unfair.

52. Silzer to Cohen, May 22, 1924.

53. These events are described in Englewood Board of Trade, letter to Silzer, April 3, 1924; and Ammann to Silzer, May 7 and May 27, 1924. For newspaper reports of Ammann's efforts during these months, see "Fort Lee Bridge Is Advocated by Engineer Ammann" (1924); "Bridge at Fort Lee Sure, Says Ammann," (1924); and "Bridge the Hudson Meeting Monday: Engineer Ammann Will Tell of Proposed Structure" (1924).

54. Ammann to Silzer, July 23, November 23, November 29, 1924; Silzer to Ammann, November 24, 1924; *Palisadian*, November 14, 1924; *Boonton Times*, November 28, 1924.

55. See *Bergen Record*, January 13, 1925; Ammann to Silzer, January 29, February 25, and March 27, 1925. See also J. W. Binder letter to Silzer, March 4, 1925, informing the governor that the "Mackay Hudson River Bridge Association" has now been formed to urge that the Port Authority build the Fort Lee bridge.

56. Silzer to Gregory, April 14, 1925.

57. Ammann to Silzer, April 17, 1925.

58. The initial Staten Island spans, the Goethals and the Outerbridge, were completed under Ammann's supervision in 1928. On the Port Authority's political insulation and Ammann's administrative abilities, see Bard (1942, Chap. 7). (The stone coverings on the George Washington Bridge towers, shown in Ammann's 1923 sketch and in his detailed plans, were omitted for reasons of cost.)

59. The Port Authority also attracted support because, as Al Smith had noted years earlier, it had the potential to pay for its projects via toll receipts and rents, and therefore without direct use of tax revenues.

60. Completed in 1927 after 8 years of political and technical problems, the Holland Tunnel between Jersey City and Canal Street in Manhattan took in millions of dollars a year in automobile and truck tolls.

61. The Depression sharply reduced the number of motor vehicles using the Port Authority's crossings in the 1930s, and traffic across the three Staten Island bridges was additionally hurt by construction of the Pulaski Skyway. Traffic over the George Washington and the three Staten Island bridges turned out to be insufficient in the early and mid-1930s to pay operating expenses and debt on the bonds. Without the toll revenue from the Holland, which maintained high traffic levels, the Port Authority would have been close to bankruptcy in the 1930s. See the analysis in Bard (1942, Chap. 8).

62. The last quotation is from "They Stand Out in a Crowd" (1934, p. 13); the others above are from "Poet in Steel" (1934, pp. 23-24).

63. Steinman and Watson (1941, pp. xv, 341). It should be noted, however, that Steinman was also a prominent bridge designer, that he worked with Lindenthal and Ammann on the Hell Gate Bridge, and that his relationship with Ammann was always competitive and perhaps at times antagonistic. See Billington (1983, pp. 141-146) for an interesting analysis of the relationship between the two men. That Steinman disregarded the drama leading to the 1925 decision might be ascribed partly to his animus toward Ammann and his success. Steinman does describe the construction process at the George Washington Bridge in some detail and with admiration (pp. 340-345), but Ammann is barely mentioned.

64. Here, in summary form, is what Binder says: In 1923, after Governors Silzer and Smith vetoed the legislation approving private tunnels, Binder himself took the initiative, studying the question of whether the Port Authority might take on the task of constructing a bridge or tunnel across the Hudson to Bergen County, Binder's home territory. In the course of his explorations, Binder visited the Port Authority's offices, where he met the staff, including "a quiet, retiring man" named Othmar Ammann, whose table "was covered with sketches of a great bridge which he hoped some time to build across the Hudson." After talking with "this quiet man who never raised his voice under any provocation," Binder concluded that "here was a master of his profession." But as to Ammann's idea for a great bridge, "no one knew anything about it," except a few "engineering societies" which Ammann had addressed. "What was needed," Binder saw, was "a campaign of education" in the region, "creating public sentiment in its favor." Binder and his allies thereupon organized a campaign, which was "opened by Senator Mackay" with a speech in October 1924. A series of meetings followed, and an association to advance the bridge project was formally organized on January 7, 1925 (Binder, 1942, pp. 182-188, see generally pp. 174-208).

Binder's recollection that he met Ammann prior to the fall of 1924 in the Port Authority's offices is difficult to credit. In the winter of 1924-1925, we know that Senator Mackay wrote to him at an office (which he had borrowed for professional work) on Fourth Avenue. In any event, the Binder-Mackay organization began long after Ammann's efforts.

65. See Bard (1942, pp. 181-185, 193). Bard goes on to describe at length Ammann's design and administrative accomplishments at the Port Authority (pp. 193-201). Bard's conclusion is that "insofar as any joint effort may be attributed to one man, the success of the Port Authority" as a construction agency "may be attributed to Othmar H. Ammann" (p. 193).

66. See the recollections of William Pallme, one of Cohen's legal aides at the Port Authority in the 1930s (letter to J. Doig, March 1988), and, generally, Doig (1988, pp. 16ff).

67. "I learned the difference between the technique of engineers and the technique of Al Smith," Cohen writes, when he looked at blueprints and other engineering work carried out at the Port agency by Ammann and his colleagues. "The engineer prepares every detail. He does not begin construction until all his plans are tested. . . . [But] if there is a statesman's job to do, another method must be evolved." Al Smith and other political leaders might have a goal in mind, but in order to win "concurrence from others—especially legislators"—they must engage in continual negotiation and compromise, an approach antithetical to that of the nonpolitical engineer (Cohen, 1946, pp. 123-124).

68. As Ammann wrote to Samuel Rea, president of the Pennsylvania Railroad, on June 12, 1923: "If this bridge proposition is not carried out the tunnel projects already underway will be pushed and supported by popular demand. . . . A popular notion, fed by tunnel advocates and . . . widely distributed sensational statements about the enormity of a bridge undertaking, appears to be gaining ground that tunnels are preferable" (quoted in Widmer, 1979a, p. 12). Rea had for several years encouraged both Lindenthal and Ammann in their bridge projects.

69. "In daytime he worked on his project, and in the evening he made speeches wherever there was an opportunity. When he returned to his Boonton home around midnight, he was tired, but still looked forward to the distraction of a game of chess with his wife. Obstinately he fought for his idea and his project. In 1924 he became a U.S. citizen, and at last, early in 1925, the States . . . gave the green light to the Port Authority . . . to build a bridge [at Fort Lee]" (Widmer, 1979b, p. 5).

70. In "Memories of My Father," Margot Ammann Durrer writes that after leaving Lindenthal, Ammann "prepared plans for and advocated construction of a more moderate bridge to cross the Hudson River between Fort Lee and upper Manhattan. . . . Father modestly glossed over the hard struggle to get the bridge under way: the bitter controversies with others of his profession, the years of working without any income, the many lectures to political groups and ladies clubs" (1979, p. 29).

71. Cohen also notes that Ammann was provided with working space—on "huge cutting room tables"—by a Boonton neighbor who was a senior official of a firm in Manhattan's garment district, and that this neighbor and Ammann's brother Ernst (who lived in Switzerland) provided some financial support in 1923-1925. And Cohen refers to Ammann's work at the Such Clay Company, which provided him with contact with George Silzer, "who would become governor of New Jersey and an influential figure in the Port Authority." The connections are not further explored (Cohen, 1987, pp. 9-12).

72. For example, Binder clearly wanted to emphasize the importance of his own role in organizing public support for the Fort Lee bridge; therefore, he had little incentive to lay out the activities of Ammann and others who worked at that task before Binder arrived. On Steinman's motivation, see Note 63.

73. For valuable discussions of engineering ideals, see Morison (1974, esp. pp. 6-8, 88-96, 127-129) and Billington (1983, esp. Chap. 1 and pp. 266ff).

74. Billington (1983, pp. 267, 269). Some of these characteristics are clearly focused on the kinds of works produced by the civil engineer. Perhaps the phrasing in Morison (1974, p. 8) captures a value that would apply to engineers more generally: "a daring elegance . . . the ultimate morality of the engineer—if it works."

75. See in particular Ammann's 1918 paper, "The Hellgate Arch Bridge over the East River in New York City," which won the ASCE prize that year for its quality of analysis and exposition; his 1931 article, "Brobdingnagian Bridges"; and his extensive 1933 report, "George Washington Bridge—General Conception and Development of Design." When teachers of English at an engineering school (Polytechnic of Brooklyn) prepared a book of essays to assist their engineering students in writing clearly, one of several essays they selected from twentieth-century engineers was another Ammann paper, his March 1926 "Tentative Report on the Hudson River Bridge" (Miller & Saidla, 1953, pp. 237-251).

76. See, for example, the studies by Herbert Kaufman, Pendleton Herring, and others, discussed in Doig and Hargrove (1987, Chap. 1).

77. To simplify the discussion, I leave aside the issue of why an extensive network of rail tunnels was not built under the Hudson River. Most New Jersey railroads terminated at the Hudson, although two did go via tunnel to Manhattan.

78. The difficulties that Ammann encountered in attempting to persuade local groups to pull together in support of the Fort Lee bridge are suggested by his letters to Governor Silzer; see, for example, his optimistic assessment of January 9, 1923, and his less happy reports of January 23, 1924 and January 29, 1925.

79. The Port Authority's commissioners were appointed by the two governors for fixed, extended terms, and the agency was expected to make its decisions on appropriate projects (for regional development) based on broad planning criteria. The fact that the Authority was expected to undertake projects only if they could be financed without recourse to tax revenues added to the agency's apparent political insulation.

80. On these three cases, see respectively Morison (1974) and two essays in Doig and Hargrove (1987): Theodore R. Marmor, "Entrepreneurship in Public Management: Wilbur Cohen and Robert Ball"; and John Milton Cooper, Jr., "Gifford Pinchot Creates a Forest Service."

81. See the relevant discussion of "imaginative reconstruction" as a crucial step in analyzing patterns of power and social outcomes in MacIver (1964, esp. pp. 258-259, 391).

82. "The relation of the biographer to the subject is the very core of the biographical enterprise. Idealization of the hero or heroine blinds the writer of lives to the meaning of the materials. Hatred or animosity does the same" (Edel, 1984, p. 14).

REFERENCES

Ammann, M. (c. 1979). *Beauty and the bridge.* (12-page unpublished paper, mimeo)

Ammann, O. H. (1918). The Hell Gate arch bridge and approaches of the New York connecting railroad over the East River. *Transactions*, 863. American Society of Civil Engineering.

Ammann, O. H. (1931). Brobdingnagian bridges. *Technology Review, 33*, 441-444.

Ammann, O. H. (1933). George Washington Bridge—General conception and development of design. *Transactions, 97*, 1-65. American Society of Civil Engineering.

A monumental bridge: New York will soon possess another "world's greatest." (1927). *Scientific American*, pp. 418-420.

Bard, E. W. (1942). *The Port of New York Authority*. New York: Columbia University Press.

Billington, D. P. (1977). History and esthetics in suspension bridges. *Journal of the Structural Division, American Society of Civil Engineers, 103*, 1655-1672.

Billington, D. P. (1979). Wilhelm Ritter: Teacher of Maillart and Ammann. *Journal of the Structual Division, American Society of Civil Engineers, 106*, 1103-1116.

Billington, D. P. (1983). *The tower and the bridge: The art of structural engineering*. New York: Basic Books.

Binder, J. (1942). *All in a lifetime*. Hackensack, NJ: Author (privately printed).

Bridge at Fort Lee sure says Ammann. (1924, April 25). *Palisadian*.

Bridge the Hudson meeting Monday: Engineer Ammann will tell of proposed structure. (1924, March 5). *Bergen Evening Record*.

Caro, R. (1974). *The power broker*. New York: Knopf.

Cohen, E., Stahl, F., & Wilson, S. (1987). *The legacy of Othmar H. Ammann*. New York: Ammann & Whitney. (Unpublished manuscript)

Cohen, J. H. (1946). *They builded better than they knew*. New York: Julian Messner.

Cooper, J. M., Jr. (1987). Gifford Pinchot creates a forest service. In J. W. Doig & E. C. Hargrove (Eds.), *Leadership and innovation* (pp. 63-95). Baltimore, MD: Johns Hopkins University Press.

Crane, H. (1966). *The complete poems of Hart Crane*. New York: Anchor.

Danielson, M. N., & Doig, J. W. (1982). *New York: The politics of urban regional development*. Berkeley: University of California Press.

Doig, J. W. (1988, September). *Entrepreneurship in government: Historical roots in the progressive era*. Paper presented at the 1988 annual meeting of the American Political Science Association, Washington, DC.

Doig, J. W. (1990). Regional conflict in the New York metropolis: The legend of Robert Moses and the power of the port authority. *Urban Studies, 27*, 201-232.

Doig, J. W., & Hargrove, E. C. (Eds.) (1987). *Leadership and innovation*. Baltimore, MD: Johns Hopkins University Press.

Durrer, M. A. (1979). Memories of my father. *Swiss American Historical Society Newsletter, 29*.

Dummies in City Contract. (1911, March 25). *The New York Times*.

Edel, L. (1984). *Writing lives*. New York: Norton.

Fort Lee bridge is advocated by engineer Ammann. (1924, March 5). *Bergen Evening Record*.

Katz, L. (1979). O. H. Ammann, master bridge builder: A remembrance. *Embassy News, 3*, 20-21.

Katz, L. (1988). A poet in steel. *Portfolio: A Quarterly Review of Trade and Transportation, 1*, 33-39.

Kull, I. S. (1930). *New Jersey: A history: Vol. III*.

Le Corbusier. (1936). A place of radiant grace. *When the cathedrals were white* (F. Hyslop & J. Finney, Trans.). New York: Reynal.

Link, A. S. (1947). *Wilson: The road to the white house.* Princeton, NJ: Princeton University Press.

MacIver, R. (1964). *Social causation.* New York: Harper & Row.

Marmor, T. R. (1987). Entrepreneurship in public management: Wilbur Cohen and Robert Ball. In J. W. Doig & E. C. Hargrove (Eds.), *Leadership and innovation* (pp. 246-281). Baltimore, MD: Johns Hopkins University Press.

MacKaye, M. (1934, June 2). Poet in steel. *The New Yorker, 10*, pp. 23-27.

Miller, W. J., & Saidla, L. E. (Eds.). (1953). *Engineers as writers.* New York: Van Nostrand.

Mollenkopf, J. H. (1983). *The contested city.* Princeton, NJ: Princeton University Press.

Morison, E. E. (1942). *Admiral Sims and the modern American navy.* New York: Russell & Russell.

Morison, E. E. (1974). *From know-how to nowhere: The development of American technology.* New York: Basic Books.

Moses, R. (1968, February 18). *Poet in steel.* Remarks on the occasion of the dedication of the Othmar Ammann College, SUNY at Stony Brook. (Available from SUNY).

Port of New York Authority. (1923). [Board of commissioners, minutes of November 21 meeting.]

Reier, S. (1977). *The bridges of New York.* New York: Quadrant Press.

Shanor, R. R. (1988). *The city that never was.* New York: Viking.

Silzer, G. S. (1923, January 16). *Inaugural address.* Trenton: State Archives, George S. Silzer files.

Silzer, G. S. [Ammann bridge file.] Trenton: New Jersey Archives.

Steel, R. (1988). Living with Walter Lippmann. In W. Zinsser (Ed.), *Extraordinary lives: The art and craft of American biography* (pp. 121-160). New York: American Heritage.

Steinman, D. B., & Watson, S. R. (1941). *Bridges and their builders.* New York: G. P. Putnam.

Stellhorn, P. A., & Birkner, M. J. (1982). *The governors of New Jersey, 1664-1974.* Trenton: New Jersey Historical Commission.

Strouse, J. (1988). The real reasons. In W. Zinsser (Ed.), *Extraordinary lives: The art and craft of American biography* (pp. 161-195). New York: American Heritage.

Stüssi, F. (1974). *Othmar H. Ammann: Sein beitrag zur entwicklung des bruckenbaus.* Basel: Birkhauser Verlag.

They stand out in a crowd. (1934, April 28). *Literary Digest*, pp. 23-24.

van der Zee, J. (1986). *The gate: The true story of the design of and construction of the Golden Gate Bridge.* New York: Simon & Schuster.

Weber, M. (1919). Politics as a vocation. In H. Gerth & C. W. Mills (Eds.), *From Max Weber: Essays in sociology* (pp. 77-128). New York: Oxford University Press.

Widmer, U. C. (1979a). *Othmar H. Ammann: 60 jahre bruckenbau.* Winterthur, Switzerland: Technorama Schweiz.

Widmer, U. C. (1979b). Othmar Hermann Ammann, 1879-1965: His way to great bridges. *Swiss American Historical Society Newsletter*, 5-6.

Wisehart, M. K. (1928). The greatest bridge in the world and the man who is building it. *The American Magazine*, pp. 34, 183-189.

3

Economic Development and Infrastructure: Planning in the Context of Progressive Politics

ROBERT MIER

In April 1992 Chicago became an international infrastructure story when a century-old system of underground tunnels sprang a leak, flooding scores of downtown buildings. As the drama played out over a several-week period, the Chicago disaster became a metaphor for a national crisis of failing public infrastructure and a failing public bureaucracy (Corral, 1992, pp. 14-19). Regardless of whether that metaphor was appropriate, it stood in stark contrast to another long-standing Chicago metaphor as the "city that works," the city of big, brawny shoulders and big public works projects as the embodiment of a vision of progress. This juxtaposition of metaphors, Chicago as failing infrastructure and Chicago as big, modern public works projects, provides a backdrop for examining the role of local government in inspiring a vision of the future and in building and managing the modern city.

Chicago generally is regarded as the prototypical American industrial era city, the city of steel mills and hog butchers, vacuum tube televisions and soup production (Warner, 1972, pp. 85-112). It also is the city of Daniel Burnham and big plans, Frank Lloyd Wright and "prairie style architecture," and big business and the first skyscrapers. As the twentieth-century archetype, Chicago was slow to move toward the twenty-first century. It was the last major city to hang on to "machine" politics and the concomitant "business as usual." This was embodied in the two-decade reign of Mayor Richard J. Daley, a period characterized by a "growth" mentality focusing on large-scale public works projects and cozy business partnerships (Green & Holli, 1987).

The Daley era saw the construction of a major freeway system and O'Hare International Airport to preserve Chicago's role as a major transportation hub; expansion of the fixed-rail public transportation system; development of the country's largest convention center; and development of three public universities and an expanded community college system. This growth era also saw the decline of the steel, meatpacking, and electronics industries; the social, economic, and physical devastation of many working-class neighborhoods; and the largest of the 1960s urban riots, a precursor of Los Angeles, 1992. The Daley years comprised a time of constant tension between a city being built and a city falling apart and set the stage for Harold Washington, the first Chicago mayor to defy the "builder" tradition (Holli & Green, 1989, pp. 139-162).

The Washington approach to development was antithetical to the traditions of the urban growth machines (Mier, Wiewel, & Alpern, 1992). Its "grass" roots were most clearly articulated in the origins of the Community Workshop on Economic Development (CWED).[1] CWED emerged from a 1982 conference sponsored by the Community Renewal Society (CRS) to critique enterprise zones, the main urban development initiative of the Reagan administration, and to heighten the local urban development policy debate.[2] The participating community organizations were frustrated with President Reagan's cutbacks and the dismantling of urban programs. Closer to home, Chicago participants were disenchanted with the continuing state and local tradition of "big ticket" public works investments as a stimulus to development paralleled with a disinvestment of Chicago's neighborhoods and manufacturing base. But they also understood that the times required strategies that moved beyond statements of opposition to enterprise zones.

The groups coalesced around the preparation of a proactive policy and program statement designed to undergird their fight for resources as well as their critique of programs at the city and state levels. They wanted their statement to reflect their experiences in delivering community development projects.

The CWED policy statement, a codification of the decisions of two statewide meetings, was written in August of 1982 (see CWED, 1982; Gills, 1991). Its seven goals emphasize the importance of neighborhoods as a social entity and their right to be authentic partners in any development process that affects them. They emphasize the importance of small-scale, targeted public works projects as development stimuli.

Finally, they reflect a perception of social justice as targeting the work needy.

Harold Washington agreed to run for mayor in late October 1982. One of his first campaign actions was to organize a participatory structure for research and issues formulation. Washington's approach to issues formulation for the campaign would characterize his approach to issues formulation in government. The approach also reflected his tendency as a legislator to hear all views. He formed issues teams to bring together diverse viewpoints and constituencies and to produce a popularly generated platform.[3] Each issue team would represent the class, race, and neighborhood diversity of the city.

By early January about 15 issues teams had started working on topics, including energy, housing, jobs, senior citizens, women, fiscal policy, transportation, and neighborhoods, among others. Each issues team produced policy papers, specific briefing papers for speeches, endorsement sessions and debates, and campaign literature. The policy papers were published in the central policy document of the campaign, *The Washington Papers* (Committee to Elect Harold Washington, 1983).

The issues team assembled for economic development included several key participants in the CWED process. Not surprisingly, the economic development issues paper, "Jobs for Chicagoans" (Committee to Elect Harold Washington, 1983, pp. 2-7), was an elaboration of the CWED platform. It articulated five main goals and eight complementary policies. Like the CWED statement, "Jobs for Chicagoans" emphasized neighborhoods, affirmative action, and targeting of work needy people, and a need for innovative, aggressive, and "appropriate" public investments. Going beyond CWED, it also emphasized job development instead of real estate development, and business retention and expansion instead of attraction.

In essence, these goals and policies represented a substantial departure from traditional, growth-oriented approaches and evoked considerable controversy (Clavel & Wiewel, 1991, pp. 1-22; Holli & Green, 1989, pp. 3-18). There were three elements to this controversy, all a reflection of the populist, participatory leanings of Washington's campaign. The strong emphasis on expanding employment (the rhetoric of "full employment" was used) evoked fears of government interference in business affairs. The emphasis on partnerships including neighborhood and working people threatened a historical cozy relationship between city hall and business and labor leadership. An emphasis on "jobs as the bottom line" threatened the real-estate-based growth coalition and the tradition of large public works projects as engines for development.

The economic development issues paper promised a departure from previous public economic development practice and set a context for strong business interest contention with the new mayoral administration of Harold Washington. This chapter will attempt to illustrate these tensions by examining three public works initiatives that confronted Mayor Washington: a proposal for Chicago to host the 1992 World's Fair; Washington's efforts to redirect public infrastructure investments; and finally, Washington's own reexamination of the role of large-scale projects in rebuilding the public city.

World's Fair

In 1981 Mayor Jane Byrne dramatically announced Chicago as the host for a World's Fair to be held in 1992. The chosen date celebrated the 100th anniversary of the famous 1892 fair and the 500th anniversary of the Columbus voyage. Mayor Byrne's announcement culminated several years of behind-the-scenes promotion, planning, and negotiating conducted by Chicago's business and civic elite.[4]

In promoting the Fair to Chicago's citizenry, Byrne and the business leadership employed rhetoric of vision, celebration, adventure, progress, self-sufficiency, and leadership. There was no clear Fair concept, and concrete facts were limited to an approximate site (along the near south lakefront within a mile and a half of the heart of downtown), cost (initially reported as $67.5 million), attendance (more than 50 million in 6 months, an average of about 300,000 per day, with peak days drawing 0.75 million), and economic benefit (almost $300 million in tax revenues alone). In the business lexicon of its day, it was hailed as a classic "win-win" opportunity to be realized through "public-private partnership" ("We Start Planning," 1982).[5]

Despite a full court media press by the Fair promoters, some of Chicago's citizenry remained skeptical. That skepticism—shared by neighborhood development activists, substantial segments of the African-American community, and opponents of Mayor Byrne—was rooted in a number of historical experiences.

First, the Fair promoters, now operating openly as "The World's Fair—1992 Corporation," were viewed as the same leaders who had advanced through the 1970s a vision of a service-based economy while treating the neighborhoods and the deteriorating industrial base with benign neglect. This vision was codified in the "Chicago 21 Plan," prepared not by the City's planning department but by a civic association dominated by businesses and downtown property owners, the Chicago Central Area Committee (1973).[6]

Second, celebrations as engines of development, as promoted by Mayor Byrne through a series of major lakefront summer festivals called *Chicagofest*, had fallen into disrepute, especially within the African-American community (Alkalimat & Gills, 1989, pp. 31-33; Gills, 1991, pp. 34-63). In fact, a boycott of the 1982 *Chicagofest* organized by Rev. Jesse Jackson and Operation Push, has been credited not only with launching the voter registration drive culminating in Harold Washington's election as mayor but also with shifting Jackson's political work into the electoral arena (Rivlin, 1992, pp. 77-79).

Finally, the Fair was the central element of the mayor's re-election campaign, and her popularity was plummeting. Her campaign media tried to build an image of a city having fun, a theme soundly rejected in the African-American community, which was experiencing growing unemployment (Rivlin, 1992, pp. 126-144; Squires, Bennett, McCourt, & Nyden, 1987, pp. 90-91).

Skepticism over the Fair quickly coalesced into a broad-based coalition of citizens' organizations demanding a "fair Fair," the "Chicago 1992 Committee." The coalition included constituency-based, grassroots organizations in white, African-American, and Latino neighborhoods as well as "good government" civic groups like the League of Women Voters and religious-based ones like the Jewish Council on Urban Affairs. Shlay and Giloth observe: "This movement was not a grassroots upsurge of rallies, civil disobedience, and folk songs against the Fair. Rather, [it was] an articulate and savvy coalition of some 47 civic, neighborhood, and environmental organizations, schooled by decades of antidowntown organizing" (Shlay & Giloth, 1987, p. 320). The committee quickly established itself as a group capable of injecting research-based facts into the public debate when it called into question the cost and revenue estimates.[7] An important comparative base for this research were the New York, Montreal, and Osaka universal category fairs, each of which had lost substantial sums of money (Shlay & Giloth, 1987, p. 308).

Despite mounting criticism, the early rhetoric promoting the Fair emphasized "self-sufficiency."[8] Fair proponents argued that the Fair would be expected to carry any costs on site, while the City would pay for any off-site infrastructure in a manner similar to its support for all other development. This latter supposition stretched public credulity, because the Fair was never intended to be comparable in scale to any other central area developments.

To bridge this contradiction, the Byrne administration, in 1982, released a "comprehensive plan" that attempted to rationalize public investment in off-site infrastructure investment as a program the City would undertake with or without the Fair (City of Chicago, 1982). This infrastructure program immediately was called into question by the Chicago 1992 Committee.

The Chicago 1992 Committee faced a formidable hurdle in raising public awareness about costs and benefits of the Fair. The major media in Chicago, lead by the two daily newspapers, were active participants in promoting the Fair. Shlay and Giloth report that almost 20% of the 61 members of the 1992 World's Fair Corporation were board members of the *Chicago Tribune* or the *Chicago Sun-Times* (Shlay & Giloth, 1987, p. 313). It took the election in April 1983 of Harold Washington as mayor to legitimize a public debate on the Fair.

As was suggested in the previous section, Washington perceived that his mandate included reexamining traditional approaches to development, including the World's Fair (Rivlin, 1992, pp. 247-248; Wiewel & Clavel, 1991, pp. 270-293). He came under immediate pressure from the city's business and civic leadership to endorse the Fair, and it became an early test of his mayoralty.

Mayor Washington had extraordinary confidence in participatory decision-making processes, as well as rich political acumen. As soon as he had his team in place, he created a broad-based World's Fair Advisory Committee to examine the issue.[9] This would prove to be one of his major devices for expanding public involvement in government decision making (Hollander, 1991, pp. 121-145).

The advisory committee issued a report in October that was guarded (World's Fair Advisory Committee, 1983). The mayor had not taken, nor would he ever take, an unequivocal position on the Fair. He already was being labeled "anti-development" and was extremely cautious about adding fuel to that criticism (Rivlin, 1992, pp. 247-248). Additionally, the mayor remained open to the possibility that, in the course of public debate, a conception of a Fair would emerge that would engender broader public support (Shlay & Giloth, 1987, p. 319). Reflecting this reticence, the committee neither endorsed nor opposed the Fair, instead raised a series of questions about it. The questions paralleled those being advanced by the Chicago 1992 Committee. They challenged the visitor and revenue projections and questioned whether the public portion of the Fair costs would be so great that other public opportunities would be foregone, and needs unmet. This sharpening debate over opportunity costs was clearly seen as a debate over infrastructure policy and priorities.

The World's Fair planners, working with the Chicago World's Fair—1992 Corporation, had projected a $25-$30 million cost for off-site infrastructure improvements. These consisted of water, sewer, and electrical service to the site. (The most expensive was a $20-million sewer main.) The advisory committee identified at least $75 million of additional infrastructure costs. Included were relocating of Lakeshore Drive; submerging a major thoroughfare in Lincoln Park to make an entrance to the fairgrounds; and rebuilding of Meigs Field, a downtown airfield within the Fair boundaries, which would be shut for a 3-to-4-year period. Additionally, they suggested enormous transportation and pedestrian improvements would be needed to funnel 750,000 people to the fairgrounds.[10] Finally, the committee observed a need for extraordinary municipal services such as police and fire protection, traffic control, and so on.

The World's Fair promoters took the advisory committee report as a rejection, and mounted enormous pressure on the mayor and the governor to endorse the Fair.[11] The governor, James Thompson, was an astute politician adept at walking a political tightrope, and in late 1983, he supported in the state legislature the creation of a World's Fair Authority to assume responsibility for planning and implementing the Fair. However, the governor's approach also was guarded. He gave the Authority limited resources pending negotiation of an intergovernmental agreement between the Authority, the City of Chicago, and the state.

Appointments to the new Authority were split between the governor and the mayor. The governor's appointments were drawn from the World's Fair—1992 Corporation. In what would plague the mayor throughout his first term, he found himself under enormous public pressure to include among his appointees some number of business people. On this as well as other boards, commissions, and corporations to which he could make appointments, the preponderance of his appointees, like Rev. George Riddick of nationally known Operation Push, had strong grass-roots and reformist leanings or were from the minority business sectors.[12] Nonetheless, he appointed several established business or civic leaders.[13] This effectively gave majority control to the Old Guard.[14]

The Authority quickly exacerbated the conflicts over the Fair by attempting to hold meetings closed to the public, effectively locking out the Chicago 1992 Committee. They also released even more inflated economic benefit estimates. The Authority now claimed the Fair would have a $2.5-billion impact on the local economy (McClory, 1986, p. 25). Despite the City's weak position with the Authority, the mayor's staff did lobby his appointees to demand an independent economic feasibility study.[15] Because of this pressure, the Authority hired an out-of-town consultant, A. D. Little, to undertake it instead of a local firm, Laventhol and Horwath, with close connections to the original World's Fair—1992 Corporation.

Under pressure from the governor and the legislature to codify an intergovernmental agreement with the Authority, the mayor released, in March 1984, a list of more than 50 conditions the agreement would have to meet. There were effectively two central demands: that the Fair carry its own costs, and that sufficient City controls be in place to guarantee the Authority would not commit the City to obligations it could not meet.

In April and May a series of a half-dozen all-day negotiation sessions was held to forge the agreement.[16] The conscious City strategy had two main thrusts: to attack the facade of self-sufficiency and to open the planning process to greater public scrutiny. The Authority commenced the negotiations by proposing to absorb the cost of all infrastructure improvements "specific to the Fair," as opposed to already planned improvements. The reference point for planned improvements was the 1992 Comprehensive Plan. This plan was regarded by the Washington administration as a weak rationalization for the Fair, not a comprehensive infrastructure plan for Chicago. Thus, much of the ensuing negotiations revolved around the issue of what was Fair-related infrastructure versus what infrastructure would be built regardless of the Fair.

Slowly but surely, more and more infrastructure was identified in the negotiations as Fair-specific. First, the Authority agreed to pay for off-site transportation improvements and water and sewer lines. Then it agreed specifically to install pedestrian bridges across railroad tracks and Lakeshore Drive into the fairgrounds. It agreed to absorb all costs associated with closing and reopening Meigs Field. It agreed to pay for new police and fire department facilities in the vicinity of the fair-grounds. No party would agree to pay for the relocation of Lakeshore Drive, so all agreed to jointly seek state or federal funding for this $30-million item. Finally, the Authority agreed to pay for 75% of all the operational costs for delivery of municipal services required by the Fair.

On the planning front, the strategy was to require all the Fair land-use decision making to be subject to plan commission review. Critics of the Fair were demanding some sort of extraordinary review process, an idea that rubbed against a basic philosophy of government held by the mayor. Washington had enormous faith in the potential efficacy of government, and focused his own reform efforts on making it work rather than circumventing it. Hence, it was important to him that government processes like the plan commission reviews be challenged to respond, rather than be circumvented.

By the time the intergovernmental agreement was complete, any notion that the World's Fair could be sponsored through a combination of private sponsorship had evaporated.[17] The Authority had agreed to absorb almost $250 million in costs that it previously assumed were either the normal support a city provides any development or were improvements already planned for the area.[18]

The City had accomplished its cost containment and process objectives and felt comfortable that realization of the Fair would financially depend on private investors and/or the state. Notwithstanding, the Chicago 1992 Committee was critical of the agreement, particularly its reliance on normal City planning reviews as a means of control, and for ambiguity about ultimate ownership of the fairgrounds at the conclusion of the Fair. The ultimate measure of the agreement, for the City, occurred quickly, when the Authority appealed to the legislature to underwrite $900 million in revenue bonds to build and operate the Fair. The legislature deferred action, but "self-sufficiency" was buried forever.

By this time, public sentiment was tilted against the Fair unless a proposal could be developed that would yield benefits better distributed into areas of the city and among populations in need.[19] The whole process had created a new level of community awareness about the impact of large public works projects and the infrastructure needs of particular communities. This awareness was sharpened by a number of quite sophisticated studies generated by the Chicago 1992 Committee, including Greer's on infrastructure and Wiewel's on employment (Shlay & Giloth, 1987, p. 319).

As a result, the next—and last—year of life for the Fair was spent with the Authority scrambling to develop a concept plan that could generate enough enthusiasm to justify public financing. The Authority shifted the conceptual emphasis from an exhibition-based one, as has been the tradition for universal category fairs, including the 1992 Seville Fair, to an entertainment-based one. An exhibition-based event is organized around pavilions sponsored by countries or corporations. An entertainment-based event is organized around sporting events, shows, and other galas. In any fair, there is a mix of the two.

With this shift of emphasis, much of their planning effort focused on major public projects, such as a domed football stadium that might serve the city and the Fair. Discussions were already under way between the city and several of its professional sports teams about new stadiums (Mier & Moe, 1991, pp. 64-99). In an effort to link such public efforts to the Fair, the Authority invented the construct of a "presidual," a Fair residual that would be available for general use before the Fair.

Given the limited political and public support for the Fair in Chicago, the Authority, with the support of the governor, attempted the classic public works "Christmas tree" strategy—promise something good all over Illinois. It developed the concept of "gateways" to the Fair in faraway nooks and crannies of the state, justification for spreading the state's investment and buying votes.

Although Illinois Governor James Thompson was a very popular Republican, the legislature was controlled by Democrats. Thompson's Democratic alter ego was House Speaker Michael Madigan. Early in 1986 Madigan appointed a 14-member advisory committee to help him sort through the mess of conflicting opinions about the Fair. The committee was chaired by former two-term U.S. Senator Adlai Stevenson III. In April the committee issued a report that effectively substantiated the positions of the Chicago 1992 Committee and the City. In essence, it said the Fair was a large financial risk and likely would divert public resources from other, more necessary projects.

In the spring of 1986 the Authority again turned to the Illinois state legislature to be the financial guarantor of the Fair. However, in May, A. D. Little released its economic impact study predicting that the Fair would lose between $50 and $350 million. Coming on the heels of the Stevenson committee's report, the Little study sounded the death knell for the Fair. On June 20, 1986, Speaker Madigan announced that the Illinois House would not consider any further action on the Fair, effectively terminating the Authority as of July 1, 1986.

The Authority was dissolved, but many of the original promoters carried on the quest for the Fair. For example, in 1987 there resurfaced the idea of a Fair as a cultural festival dispersed among Chicago's many museums and theaters. By this time, the Washington administration had undertaken two neighborhood-focused infrastructure bond issues, and the popular climate for festivals as an engine of development had waned.

The Neighborhood Infrastructure Program

The debate over the World's Fair and the 1992 Comprehensive Plan had heightened community awareness about wide-ranging infrastructure needs around the city. In a series of neighborhood meetings held monthly, the mayor began to hear people articulate their needs in infrastructure terms. They would speak eloquently about collapsing vaulted sidewalks, streets full of potholes, clogged sewers, deteriorating alleys, undermaintained or nonexistent branch libraries, and overcrowded municipal facilities such as health and social service centers. They were concerned about jobs and income and wanted the mayor to be their champion on these issues. But they also saw the connection between the basic physical quality of their neighborhoods and their social well-being. Although they loudly rejected the World's Fair as a means to build and rebuild the physical environment of the city, they nonetheless expected the mayor to act.

As was observed in the introduction, Washington, as a candidate, had articulated a planning and development vision with a neighborhood focus (Mier, Moe, & Sherr, 1986, pp. 299-309). This was further articulated in his *1984 Development Plan* (Krumholz, Kostigan, & Keating, 1985, pp. 395-396). When viewed against more mainstream approaches to development, its uniqueness was clearly evident.[20] His plan emphasized job development more than real estate development; neighborhood development instead of central business district development; retention and expansion over attraction of industry; and small businesses over large. Additionally, it consciously and overtly sought to target the work needy instead of relying on the labor market to distribute jobs. Finally, it sought broader partnerships involving community and civic actors (Krumholz et al., 1985).

From the perspectives of his *1984 Development Plan* and the neighborhood forums, it was easy to see Washington's interest in developing a neighborhood public works program. His main obstacles would be resources and politics.

Washington had inherited a $100-million operating budget deficit when he took office in 1983. Not surprisingly, considerable early focus on his administration concentrated on his fiscal capability. By the end of his first year in office, he had arrested the deficit through a combination of personnel cuts and tax increases. With the operating budget under control, he began to look at the possibility of a capital improvements program.

This appealed to him for both fiscal and political reasons. Fiscally, his financial advisors proposed a combination of retirement and refinancing of older debt, especially debt incurred in the late 1970s when interest rates had soared. If this was done, they predicted he could borrow for traditional public works uses, without a tax increase, an additional $100 to $150 million. Washington knew this move would draw political opposition, but he suspected strong community support based on what he heard at the neighborhood forums to be strong citizen concern about deteriorating local infrastructure.

In late 1984, as he was cooling toward the World's Fair, Washington directed his capital improvement team—the planning, publics works, and infrastructure directors—to prepare a $120-million program with primary emphasis on neighborhood infrastructure.[21] The planners turned to the City's most recent Five Year Capital Improvements Plan as a first source of projects. Within about one month, a proposal was presented to the mayor for review.

Washington had created in 1984 a small policy advisory council and used this body to review the proposal.[22] To the surprise of the advisory council, the proposed infrastructure program was substantially skewed toward wards represented by the 29 aldermen aligned in political opposition to the mayor. The program was rejected, and the capital improvements team was directed to present an alternative with a more equal distribution of projects across the city (Hollander, 1991, pp. 127-128). This was not hard to accomplish: Although the poorer African-American and Latino areas had suffered years of neglect, even the more affluent northwest and southwest sides of the city still had large concentrations of Depression Era, substandard streets constructed through the Works Progress Administration.

To digress for a moment, there were a number of hypotheses as to how the original infrastructure proposal, a politically naive and dangerous one, could have been advanced to the mayor. They ranged from political ineptitude to conscious conspiracy. Those interpretations emphasizing ineptitude challenged claims of professionalism on the part of the capital improvements team. Those emphasizing conspiracy believed that the mayor's political opposition controlled vast reaches of the bureaucracy, were determined to continue the pork barrel method of allocating public works projects, and the commissioners in charge of the infrastructure departments were not able to prevent it.[23]

After a careful examination, two forces seemed at work, neither of which was incompetence, pork barrel, or conspiracy.[24] First was the dominance of technicians in the planning process; and second was the organizational culture that generated projects. Technicians are quick to claim jurisdiction over health and safety issues, and often claim that infrastructure priorities ought to reflect an orderly attack on these concerns.[25] It was difficult in planning the infrastructure bond issue for nonengineers and architects to challenge these claims. Therefore, other policy priorities, such as a desire to stimulate private investment, took a backseat.

The impact of organizational culture transcends pork barrel. Most of the infrastructure needs of the city were repairs and replacements—of sewers, sidewalks, streets, bridges, and the like. Short of a complete breakdown, these projects most often entered the pipeline based on the report of a field inspector. (Complete breakdowns were the second most common source.) The field inspectors, on the whole, had a long tenure with the City. This meant that they were most likely to be white, politically aligned with the Democratic machine, and inhabitants of the city's southwest or northwest sides. And most problematical, they were likely to observe a less serious infrastructure need in one of their neighborhoods as a higher priority than a more serious infrastructure need in a minority neighborhood.[26] Technical evaluation of infrastructure is essentially a judgment call, and the social and cultural attitudes of the technicians played a prominent role.

Whether the explanation was incompetence, pork barrel, conspiracy, technical dominance, or organizational culture, the original package of projects for the infrastructure bond issue was rejected by the mayor and his policy council. Within about 2 weeks, the capital improvements team presented the mayor an alternative with a more uniform distribution of projects across the city. The key to this was a commitment to resurface five miles of residential streets in each ward at a total cost of $50 million.

Of additional significance was the addition of almost $20 million of infrastructure supporting commercial and industrial areas of the city. In Chicago, City-supported infrastructure programs had ignored these areas because they were perceived as lacking a political constituency.[27] Yet the mayor insisted on their inclusion because he sensed from his visits to community forums that Chicago citizens, in the course of undertaking more community planning, were developing a keen sense of the economic interconnects in their communities.

The mayor submitted this proposed package of projects as a bond issue to the City Council in early 1985. Despite the obvious benefit of the capital improvement program to the entire city, the proposal languished in City Council through the spring. The rationale of the mayor's opposition in stalling the program was eloquently and cruelly stated by Alderman Richard Mell. Mell said:

> There are some who believe that to get rid of Harold Washington is good government because we simply can't take four more years of him. Maybe . . . two years of not having this [bond] issue is worth ten years of political stability in this city.[28]

In the face of such staunch opposition, the mayor decided to take the issue to the streets (Reardon, 1990, pp. 176-193). Using a bus filled with reporters from print and electronic media, the mayor began to visit wards of key opposition aldermen. His goal was to shift four votes into the ranks of supporters of the infrastructure program. The mayor literally walked the streets, rang doorbells, and, with cameras rolling, asked residents for support to fix up their neighborhood.[29]

The opposition aldermen, especially in the wards visited by the mayor, came under intense popular pressure. Citizens clearly understood that a comprehensive neighborhood infrastructure program was within reach, and they would not permit it to get bogged down in partisan politics. Notwithstanding, the ultimate cost of the four swing votes was an enormous expansion of the program, ultimately to more than $160 million. Some of the growth was in citywide projects like preliminary repairs to historic Navy Pier on Chicago's lakefront. Most of it was in commitments to rebuild antiquated residential streets on the southwest and northwest sides of the city. These streets originally had been Depression Era Works Progress Administration projects and did not meet City street code standards. Needless to say, these areas of the city constituted the heart of the mayor's opposition. Table 3.1 delineates the entire bond issue.

Passage of the neighborhood infrastructure program was only half the battle. Getting it built was the other half. The building challenge was compounded by the mayor's commitment, articulated in his *1984 Development Plan*, to open up the business of the City to contractors who had been excluded (City of Chicago, 1984).

TABLE 3.1 Proposed Public Infrastructure Bond Issue
Mayor H. Washington, 1985

Residential Street Resurfacing	$40,560,000
Sewer Installations	27,750,000
New Street Construction (WPA Streets)	22,751,500
Renovation of Navy Pier	18,000,000
Industrial and Business Area Improvements	14,775,000
Vaulted Sidewalk Reconstruction	11,100,000
Municipal Office Building Repair	9,590,000
Replacement of Traffic Signals	5,000,000
Residential Sidewalks	3,000,000
Viaduct Reconstruction	2,104,000
Centerline Drainage	2,000,000
Street Lightpole Replacement	1,400,000
Residential Alley Construction	1,300,000
Street and Alley Lights	1,000,000
Repair of Firefighting Facilities	950,000
Repair of Police Facilities	940,000
Replacement of Trolley Poles	350,000
Lakefront Protection	200,000
Replace Lane Control Signals	50,000
Total	$185,000,000

SOURCE: Unpublished working document, City of Chicago, Office of Budget and Management (n.d.).

Among the reasons for undertaking an infrastructure program comprised of diverse, small projects instead of mega-projects is that they afford new opportunities for small, local businesses. For example, this concretely meant soliciting bids for 50 small street resurfacing jobs instead of several large ones that only a couple of contractors had the capacity to undertake. This increased exponentially the steps from engineering through inspection in the contracting and building process, placing an extraordinary burden on the public works and purchasing bureaucracies. It also increased the opportunity for mistakes as well as conscious impedance.

The street resurfacing example also illustrates the possibilities of impedance or sabotage. At the bid opening in late 1985 for the 50 resurfacing jobs, it turned out that the few traditional contractors were the only bidders. A meeting was held between the mayor's chief of staff and the purchasing agent to explore what might have happened.[30] The purchasing agent tried to provide assurance that he had personally assembled a list of almost 40 contractors capable of small-scale resurfacing jobs. None of them bid. As the story unfolded, it became apparent that he had no idea whether the new, potential bidders had actually received bid solicitations, even though he had "directed them mailed." Random follow-up with the contractors confirmed that *none* of them had received the solicitations. In hindsight, the best guess is that the bid solicitations were mailed at a convenient trash bin.

Such episodes notwithstanding, the administration of Mayor Washington had succeeded by 1987 in shifting more than 20% of Chicago's purchasing from non-Chicago-based firms to Chicago-based ones, by awarding bid discounts based on the concept of "net least cost purchasing." This approach recognized that Chicago-based firms were local taxpayers, and returned some amount of each contract to local revenue coffers. In total, the 20% shift meant a greater than $100-million *per year* impact on the local economy. There are few development opportunities that have the potential for such impact.[31]

Yet the mayor's development philosophy of small, widely dispersed projects with lots of opportunities for community involvement was developing some blemishes. Despite its broad popular appeal, it presented two problems. First, there were some large-scale projects that needed attention. Washington was in somewhat of a philosophical and political bind in justifying to his core constituency support for these projects.

Second, besides physically rebuilding their neighborhoods, Chicago's citizens wanted a vision of the future. As such a vision, the metaphor of "small is beautiful" failed. The mayor was becoming increasingly frustrated with the narrowness of his own presentations to citizens' groups, feeling he was reading lists of projects and accomplishments. He himself was having difficulty distinguishing the forest from the trees.

As a result, a series of "mid-course correction" meetings occurred during late 1985 and early 1986, to undertake a critical assessment of the important needs of the city. At one such meeting, a formal retreat of the mayor's entire cabinet, large-scale public works projects emerged as a priority. In part, they developed as a political imperative: the Cabinet agreed that the Washington administration, against the background of its rejection of the World's Fair, had to demonstrate its capacity to effectively implement large-scale development projects. In part, these projects were perceived to be, of necessity, an integral manifestation of a larger vision of what Chicago could be.

These arguments propelled a reexamination of the entire approach to large-scale projects. To a degree, Mayor Washington undertook this task reluctantly. The large projects risked activating the urban growth coalitions, and having them again run roughshod over neighborhood interests. But the politics of the moment demanded a return to big project public works.

Large Scale Projects[32]

Several development initiatives were emerging as issues at a scale and importance that transcended the capacity of any one City department to direct. They included the renovation or replacement of professional sports stadiums, the construction of a new central library, the development of new transit stations, and the development of surplus land on the City's airports. None of these projects was of central concern to the mayor's core constituency, and each generated controversy. Neighborhood groups' views about the correct development course were clearly defined by their own interests. They tended not to prioritize projects with a citywide impact or benefit (Brehm, 1991, pp. 238-269). Harold Washington's track record on major projects was not their concern. Two projects, in particular, underscore the dilemmas faced by the Washington administration as it returned, almost inexorably, to a policy of large-scale public works. They were the stadium projects and the central library.

The White Sox

Three different stadium projects surfaced during Mayor Washington's tenure. Both the Bears and the White Sox wanted new stadiums, so initial efforts were focused on finding a site that could accommodate a new stadium complex. After some early consideration of a domed stadium, it became apparent that neither team wanted to play indoors or on artificial surfaces. Further, they had vastly different needs in terms of stadium size, with football requiring a 75,000-seat stadium, and baseball a 45,000-seat one.

By July of 1986 any prospects for a stadium complex with two teams as tenants evaporated, and the White Sox announced that they were going to leave Chicago to play in west suburban Addison, Illinois. Needless to say, the city was in an uproar. Soon the battlecry to "Save Our Sox" was dominated by a chorus of activists who were more interested in baseball than they were in the neighborhoods around Comiskey Park. Their organizing efforts quickly were complemented by political activists seeking to build an organizational base in the communities adjoining the baseball park. They focused their attention narrowly, on keeping the White Sox in Chicago and retaining historic Comiskey Park. They built grass-roots support among White Sox fans. They did not really entertain the possibility that keeping the White Sox in Chicago might mean building a new stadium for them.

More important, the Washington administration failed to expand the issue into a broad community context.[33] This failure had multiple roots. This was the first of the mayor's big projects with a direct community impact, and it had emerged in the midst of the rejection of the World's Fair and the concomitant advancement of a neighborhood public works agenda. Not surprisingly, the mayor was tentative in embracing it and reluctant to raise neighborhood hackles.

In addition to this policy ambiguity, there were administrative impediments. The project did not really fit in any department. As such, the ad hoc team appointed to negotiate with the White Sox and the state was always trying to borrow resources from other departments. City departments responded by trying to minimize staff effort, in both quantity and quality.

By mid-fall of 1986 support for the White Sox in Addison had eroded, and in early November their bid to move was defeated in a local referendum by a close vote. Meanwhile, the White Sox were being quietly courted by St. Petersburg, Florida, and Denver, Colorado. The White Sox were ambivalent about leaving the Chicago area and, after 2 weeks of "licking their wounds," presented the City with another opportunity to "Save Our Sox."

The Illinois state legislature was due to convene for a brief session in late November, on the Monday following Thanksgiving. After several days of shuttle diplomacy between the City and the White Sox, the mayor decided, on Thanksgiving Day, to try to forge an agreement to present to the Assembly. Given the rapid pace of negotiations and his desire to "surprise" the Assembly with a proposal, he felt he could not launch an a priori community process. Nonetheless, he believed he was simply executing a community mandate.

Things moved very quickly. In a matter of 48 hours after Thanksgiving, an agreement between the City and the White Sox was negotiated and signed by the mayor and the team owners. Pending approval of the State Assembly, the deal would keep the White Sox in Chicago if the state would build a new stadium in the vicinity of the existing Sox stadium. That location choice, which would prove controversial, was totally driven by costs—reusing existing infrastructure would save

$30-50 million. Within 5 working days, it was passed by the General Assembly. The financial structure was crucial to the mayor: The new stadium would be financed by rent paid by the team, supplemented with a new tax on hotel and motel rooms. The rationale for this tax base rested with the argument that it would not be available for any local uses other than those seen as benefiting the "visitor industry".[34]

In the short run, there was widespread euphoria. After Mayor Washington's death, much controversy erupted, which warrants exposition elsewhere. In a nutshell, the White Sox walked away from the deal, raised the stakes, and forced another $150 million of public subsidy into the deal. Also, the Illinois Sports Facilities Authority, empowered with building the stadium, was accused of running roughshod over the community, and in so doing, exposed the lack of roots in the project's original neighborhood of the community organizing and City planning efforts (Mier, 1991a, p. 1).

The Bears

After the collapse of the domed stadium idea, and while the courtship with the White Sox was unfolding, there emerged a focus on the city's football team, the Bears, and its home, Soldier Field. Two forces shaped the direction that deliberations with the Bears took. The Chicago Central Area Committee became strong advocates for a privately financed stadium, and stood poised to organize the business community to purchase the luxury seating that could make such a venture feasible.[35] At the same time, a west-side, grass-roots organization, the Midwest Community Council, marshaled a campaign to bring any new Bears stadium to their neighborhood (McCullom, 1992, p. 43).

Mayor Washington was initially reluctant to consider the west side. Although there was considerable vacant land in the area being advocated by the Midwest Community Council, there also remained a significant number of occupied housing units. In late 1986 he proposed a site immediately north of the existing Bears lakefront stadium, Soldier Field. To complement the privately financed stadium, he proposed demolition of Soldier Field, moving parking off the lakefront to open up the space, and creation of a museum complex to segregate football fans from visitors to the Museum of Natural History, the aquarium, and the planetarium immediately north of Soldier Field.

His proposal was immediately scorned by lakefront protection interests, and it quickly became a major mayoral campaign issue. Their arguments captured the support of the *Chicago Tribune*, and the mayor reluctantly backed down. In January 1987 the mayor announced the creation of a site location committee in order to deflect the issue until after the election.

The mayor remained reluctant to include the west-side site in the committee's deliberations, but a strong group of organizations in that area demanded its consideration. In mid-1987 the Committee recommended the west-side site. In the course of the committee's deliberations, strong local opposition to the site also emerged. The opposition was rooted in a fledgling, church-based, community organization, the Interfaith Organizing Project.

The City's approach to the conflict between the committee and the organization of neighborhood interests opposed to the site was far different from its approach to the White Sox stadium project. This time the mayor proceeded far more cautiously. He appointed a project management team to design and implement a broad community-planning process for the Bears project.[36] Given the magnitude of the relocation problem that would occur with the west-side site, the planning process encouraged community debate. The committee, working with proponents and opponents, facilitated a process of community dialogue. A continual effort was made to reconcile community interests.[37]

As the community dialogue ensued, the cost of the community demands, although quite reasonable, began to mount. The mayor pledged to address the community needs, and decided to subsidize community improvements with tax revenues from the luxury seating that would exist in a new stadium. He saw himself undertaking an interconnected community development effort rather than a project development one, with the more substantial components of the development subsidizing the less substantial ones. The mayor believed this approach was consistent with his development policies of balanced growth and linkages between large and small projects.

In the end, the City proposed guaranteeing dislocating households the option of physically moving and improving their home, or building a new one of greater value. Further, it proposed holding these residents harmless for any increased costs they might encounter, such as increased taxes. It proposed a number of community facilities, like a library and a park. Most of this would be paid from tax revenues derived from luxury seating.

This novel approach to the use of "tax increment financing," employing amusement tax revenues instead of the usual property or sales taxes, promised to work to the advantage of the more needy. It also required amendment to the state enabling laws and shifted the decision-making locus to the State Assembly. In the meantime, Mayor Washington had passed away, and Chicago lacked the same sort of powerful leadership in the state capital. The development ran aground.

The conservative head of the Republican assembly, Pate Phillips, gave the best testimony on the Bears deal. He looked at it and said "This deal is dangerous. If we pass this it's going to set up a precedent that we can't live with elsewhere." The Illinois General Assembly rejected the deal (see Mier, Giloth, & Moe, et al., 1993, p. 105).

The Cubs

A 1982 proposal by the Chicago Cubs to play night baseball had aroused a firestorm of controversy in the neighborhood surrounding Wrigley Field. Most of it centered on the potential loss of what was argued to be a delicate balance between a high attendance density land use in a dense residential neighborhood. In response, the well-organized, highly educated, and articulate upper-middle-income community, organized as a group named C.U.B.S. (Citizens United for Baseball in the Sunshine), had engineered local and state legislation prohibiting lights.[38] Apparently fed up with the controversy, in 1984 the Cubs joined the chorus of dissident teams and proposed moving to a new stadium in the suburbs.

Mayor Washington agreed to take up the lights issue. He was partially motivated by a reluctance to avoid a thorny sports team issue affecting an upper-income white neighborhood when he was facing up to similar issues in poorer, largely black neighborhoods. Also, he saw the issue as a classic land use conflict with strong similarities to ones he faced in neighborhoods in which residential gentrification was closing in on industry (Mier, 1989, pp. 169-174).

He directed that an open, public process be undertaken to find a way to partially accommodate the Cubs, enough to make it difficult to leave the city. The process consisted of creating a negotiating committee comprised of neighborhood residents, business leaders, and the Cubs. Their meetings were professionally facilitated, supported by considerable information gathering—including the use of survey sampling—and a number of community meetings.

A compromise resulted, hammered out over an almost 2-year time period. The team was allowed to install lights at Wrigley Field. However, severe limitations were placed on the number of night games (18) and the specific starting times for games. A curfew was invoked for alcohol sales, as well as severe neighborhood parking restrictions. By all accounts, including those of the main "no lights" advocates, the solution has worked quite well.

Chicago Public Library

The construction of a new central library had been part of the platform of more than one previous mayor; however, past plans for the project had never been realized. When he arrived in office, Harold Washington inherited the library proposal from his predecessor, Mayor Jane Byrne, who had proposed putting the new facility in a renovated department store on State Street—Goldblatt's. This site had been the subject of intense controversy. Some criticized the building for its structural inadequacies, arguing that it would never be able to carry the floor weight of library books. Others raised aesthetic criticisms. Still others criticized the selection process, suggesting that the building had been chosen in the classical political "smoke-filled room," with no apparent appraisal or consultation with the library board.

Yet this controversial decision was not reversed early in the Washington administration, even though Washington had criticized the site selection process in his campaign. Initial engineering and architectural studies were started in the Department of Public Works, and within the library bureaucracy, so that the library project moved on to the mayor's development agenda by late 1985.

This change in emphasis—making the library part of the development agenda of Chicago—was due in large part to a campaign initiated by the *Chicago Sun Times*, the Better Government Association, and the Union League Club to scuttle the Goldblatt's site. The rallying metaphor became "world-class library," and provided fuel for the argument that the Goldblatt's site would both be exorbitant in expense and never get the city the library it deserved.

The mayor established an internal committee to address the questions of the site—Goldblatt's or not—and the question of financing. The committee's job was to formulate a total implementation program. The committee included Economic Development, Planning, Library, Budget, Comptroller, and Public Works. The initial recommendation favored the Goldblatt's site mainly because of cost considerations and a concern for preservation of the Goldblatt's edifice as adaptive reuse of a historically important structure. The committee felt this would be a significant statement of a development objective (Hollander, 1991, pp. 140-143).

Opposition continued, however, including a *Sun Times* front-page editorial and the intervention of Roman Catholic Cardinal Joseph Bernadin.[39] Facing this, and internal disagreement within the committee, the mayor rejected the recommendation and directed the committee and library board to work together to find a new site and construction program for a new library. After consideration of several North and South Loop sites, the library board chose a site at State and Congress streets in the South Loop.

Because of the time pressure to deliver this project, and in light of the controversy surrounding cost overruns at such other projects as the State of Illinois Center and McCormick Place, the mayor was searching for an approach that would deliver this project on time and within the budget. It was also more subtle: The mayor's team now understood the way large projects were done to enhance the functioning of traditional development networks, with featherbedding of consultant costs, inflated change orders, and so on. The mayor wanted to change this process. There was also a desire to change the process in a different way, by including a greater degree of public involvement.

The mayor established a Central Library Advisory Committee to advise him on the specifics of the development approach, in particular, completing the library as a "design-build" project. Using this technique, bids would be solicited to design and build the library within a specific cost. The developer would be selected based on both design and cost criteria. And significantly, the developer would be selected by a committee following extensive public debate on the designs themselves. In this case the designs of the competing teams were on display at the cultural center and were the subject of extensive public hearings.

This approach was a fundamentally important departure from traditional approaches to development of such projects, not only in Chicago but in the nation. It was so significant that it was the subject of a *NOVA* public television program titled "Design Wars." In choosing the design-build approach, the mayor clearly understood that he was authorizing a process that was completely insulated from political influence in the development process, even influence that might work to his advantage.

The process worked far better than he imagined. Thirty thousand people viewed the design entries at the Cultural Center. In effect a constituency for the new central library was developed through the process. Construction began after the mayor's death and was completed in late 1991. The library is named in his honor.

■ Conclusion

On the whole, I am comfortable leaving the discussion at this point and letting the stories speak for themselves. I do not pretend that the stories are "objective" and "scientific"—in fact, I reject that notion, in part because I was a key actor in the stories.[40] I think it is both fair and necessary, then, to share those appreciative elements that shaped my storytelling.[41] The predominant ones were my regards for progressive planning and partnerships.

The subtitle of this chapter is "Planning in the Context of *Progressive* Politics." The notion of "progressive" planning and politics is best elaborated in the work of Clavel (see Clavel, 1986; also Goldsmith & Blakely, 1992, pp. 172-192). Clavel defines "progressive" as "an approach to government that harken(s) back to early-twentieth-century reform movements which also combine the radical redistributive demands of the 'have-nots' with a pragmatic acceptance of the necessities of governance imposed by the larger array of social classes" (Clavel & Wiewel, 1991, p. 2). One way I have tried to present the three cases on Chicago is to illustrate the tension among the reform, redistributive, and pragmatic dimensions of progressive planning.

The reform dimension is illustrated, for example, by a variety of efforts to shape public accountability of government, often against the desire of the traditional growth coalitions for a more exclusive process. These reform efforts include the Chicago 1992 Committee's quest to gain access to World's Fair planning information and to open the decision to broader public debate; by the design-build process of the library; and by the continual struggle to expose the true costs of public works projects.

But the efforts for accountability also are shown to clash with another reform tendency, "professionalization" of public decision making. The early stages of planning for the infrastructure bond issue and the planning for the White Sox park are examples of efforts to illustrate this tension and provide entrée for redistributive claims. Ultimately, it is the juxtaposition of the neighborhood infrastructure bond issue and the World's Fair that best illuminates this tension.

It is telling that Mayor Washington returned to big projects, albeit uncomfortably. He seemed to find in these projects a pragmatic vehicle for stimulating public discourse about his vision of social justice and equity, a discourse that often got lost in a forest of small projects. As I have discussed elsewhere with Joan Fitzgerald, it is this difficulty of provoking visionary discourse that frames a second approach to my storytelling, the emphasis on partnerships (Mier & Fitzgerald, 1991, pp. 268-279).

For a local development agent advancing a project, whether working for the City, the local growth coalition, or community organizations, the rough-and-tumble work, occurring in both boardrooms and barrooms, is developing consensus, forging broader alliances, fighting rearguard actions, and finding a confidence-building public language. The popular vehicle for consensus building around "visionary" public works projects like Chicago's World's Fair is the public-private partnership.

On close examination, the concept of partnership turns out to be quite ambiguous. First of all, it is not a scientific category, firmly bounded and standardized. Rather, it is a generative metaphor. Partnership connotes, for example, the image of a community's leadership rowing in a boat together, sometimes against the strong tide of economic restructuring, and introduces strategies such as codevelopment.[42]

Marris, Morgan, Schön, and Lakoff and Johnson say generative metaphors are essential to the integration of understanding and action, because they liberate the imagination and engender new understandings of problems and approaches to their solution (Lakoff & Johnson, 1980; Marris, 1987; Morgan, 1987; Schön, 1979). And yet, in Chicago we see that the partnerships metaphor sparked very different imaginations. Organizations as diverse as the Chicago 1992 Committee and the Citizens United for Baseball in the Sunshine raised basic questions not only about who gets a seat in the boat and the course it is to follow, but more fundamentally whether it is the right boat. This conflicting understanding exists because generative metaphors are partially molded by a set of relative constants that professionals bring to their practice.

Schön identifies these constants as "appreciative systems," "overarching theories," "role frames," and "media, language, and repertoires" (Schön, 1979, p. 270). The dominant set is one Marris refers to as a corporatist paradigm (Schön, 1979, pp. 128-133). Its essential features are a view of government's role as secondary to that of business and the economy, "regulating social expectations in accordance with the requirements of the economy." Yet it also sees government trying to be businesslike, and in so doing, acting more entrepreneurial in its efforts to achieve economic growth. The view draws more on organizational than on economic doctrines:

> The primary task of government is to make the most of the land, labor, skills, raw materials, infrastructures and social amenities within its jurisdiction. The intelligence of government, therefore, is likely to depend above all on how well it understands corporate behaviors, the thinking and information on which it is based. (Schön, 1979, pp. 130-131)

This corporatist paradigm shines through in the story of the planning for the World's Fair.

A second view of practice stands in contrast to the dominant corporatist paradigm. It is called by Marris an ecological paradigm, and by Friedmann and Weaver a territorial one (Friedmann & Weaver, 1979, pp. 89-113; Marris, 1987, pp. 133-138). This paradigm has its roots in the concept of natural resource limits, but it also is influenced by the experiences of community action and development. Its essential feature is a respect for the right of each community of people to a "familiar" habitat, hence a recognition that policy planning cannot be decided apart from a particular context. The paradigm emphasizes: "Social responsibility against economic autonomy; decentralized, democratic control against remote, concentrated, corporate hierarchies of control; and understanding the whole against the abstraction of partial relationships" (Marris, 1987, p. 137). This seems to capture the spirit of the World's Fair opposition and the neighborhood infrastructure bond issue and is manifest in some of the latter large-scale projects.

Even where acknowledging the potential of this frame of reference, planners encounter political barriers that prevent them from effectively embracing it. It demands that public planning be directed toward explicating complex patterns of interrelationships. Marris believes this would require a politics capable of inspiration by generative metaphors more ecological in nature, ones capable of stimulating the process through which groups with disparate interests arrive at a mutual agenda.[43]

I have tried to describe partnership formation as a process of building and sustaining fragile relationships in new types of formal and informal organizations. Effective development relationships must face such risks as being open to redefining and resetting the problem.[44] And as they strive to maintain harmony, they must be capable of transformation and invention of new patterns of behavior.

NOTES

1. There were many roots converging: community organization experiences of the 1970s, the movement for black political empowerment, the behavior of Chicago's traditional growth coalition, community conflict with former Mayor Jane Byrne, alienation in the small and minority business sectors, and the impact of the series of recessions beginning in 1973. See Clavel and Wiewel (1991) and Alkalimat and Gills (1989, pp. 31-33).

2. Community Renewal Society (CRS) is a civic organization sponsored by the United Church of Christ. Its mission throughout its 100-year history in Chicago had been the care for and development of low-income communities.

3. In choosing this course, he rejected the more traditional alternative of assembling a few "experts" for a short, intensive effort to shape the campaign issues.

4. See McClory (1986) and Shlay and Giloth (1987, pp. 305-324). Shlay and Giloth submit a classic sociological analysis of interlocking institutional memberships to present a mosaic of corporations, civic associations, and private clubs whose participants constituted the fair advocacy network. After analyzing memberships of more than 1,000 individuals in more than 2,700 public and private organizations, they identify a core of 17 private firms, 6 civic associations, and 4 private clubs that constitute what they call the "central" members of the pro-growth coalition. This central group is dominated by Chicago-headquartered corporations, traditional growth coalition members (newspapers, banks, and utilities), and large firms with a history of civic activism.

5. For a discussion of similar public-private leadership models, see Judd and Parkinson (1990, pp. 295-307).

6. Chicago Central Area Committee (1973). For discussions about the political importance of this civic plan, see Clavel and Wiewel (1991).

7. These research efforts were assisted by scholars at local universities. For example, see Wiewel (1983) and Greer (1984).

8. The Chicago World's Fair—1992 Corporation included considerable media connections, including direct membership of the City's leading newspaper, the *Chicago Tribune*. By late 1981 the corporation was releasing a regular report called *Chicago '92 Report*.

9. In August 1983, I was appointed Commissioner of Economic Development and Chair of the mayor's development subcabinet, a policy development body comprised of the Commissioners of Planning, Housing, Employment and Training, Economic Development, and Cultural Affairs. Until late 1984, when the subcabinet model of policy development was more widely implemented within the Washington administration, the development subcabinet also included the Commissioners of Public Works, Aviation, and Neighborhoods. Several of the development subcabinet members, including myself, sat on the advisory committee, and our staffs provided staff support to the committee.

10. By comparison, fewer than 500,000 commute to work in Chicago's Loop each day.

11. In my experience, the "booster" mentality treats anything short of complete acceptance, or acquiescence, as a rejection. Thus, failure to endorse the Fair was seen by the Fair promoters as confrontation, and they responded accordingly.

12. Minority business representatives often found themselves in a complex situation. They sought and supported a progressive development agenda, yet remained very dependent on the majority business community. See Mier, Fitzgerald, and Randolph (1993).

13. Rivlin (1992) argues that members of Chicago's civic and business community who identified themselves as liberal were much more comfortable with traditional growth coalition politics than the progressive variety advanced by Harold Washington.

14. Washington subsequently would repeat this mistake with the boards governing the McCormick Place exposition center, and the Illinois Sports Facilities Authority governing the new Comiskey Park. In his first term, he effectively gained control only of the Chicago Park District.

15. A member of the Mayor's Office of Intergovernmental Relations worked closely with mayoral appointees to boards, commissions, and corporations.

16. The City was represented by its Chief of Staff, the late Bill Ware, its Corporation Council, James Montgomery, and the author. The World's Fair Authority was represented by one of its members, Donald Petkus of Commonwealth Edison Corporation, and an attorney, Jack Guthman of Sidley and Austin. The state was represented by the governor's Chief of Transportation, John Kramer, assisted by an attorney, the late Rebecca Schneiderman. Draft agreements, codified at the conclusion of each session, reside in the Harold Washington Development Archives at the Chicago Historical Society. As an aside, at the conclusion of the negotiations, Kramer would resign his state post and assume the role of Executive Director of the World's Fair Authority.

17. The final agreement, dated May 31, 1984, is titled "Intergovernmental Agreement Among the State of Illinois, the City of Chicago, and the Chicago World's Fair 1992 Authority for the 1992 Chicago World's Fair." It also is available at the Chicago Historical Society.

18. The $250-million figure was never exactly priced but was the rough City estimate developed during negotiations and corroborated in a private discussion between the author and John Kramer, the governor's representative to the negotiations and subsequently Executive Director of the World's Fair Authority.

19. Shlay and Giloth (1987, p. 319) postulate this is one major measure of the ultimate failure of the World's Fair leadership.

20. For a good discussion of mainstream approaches, see Luke, Ventriss, Reed, and Reed (1987). For comments on these approaches, see Mier and Fitzgerald (1991, pp. 268-279). For discussion of the uniqueness of the *1984 Development Plan*, see Krumholz (1991, pp. 291-300).

21. These were the individuals who traditionally prepared the annual 5-Year Capital Improvements Plan.

22. This 10-person body, which included three mayoral confidantes from outside government, was selected not by role but by personal ability to collectively bring to the table a complete perspective of government and the city. Out of a 40-person cabinet of department heads, only 2 served on the advisory council—the corporation counsel and the author.

23. These included the departments of Aviation, Public Works, Water, Sewers, and Streets and Sanitation.

24. I am a registered professional engineer and have considerable experience in construction and public works management. Because I was the only member of the mayor's policy council with this type of experience and orientation, I took a personal interest in exploring what had occurred. I conducted a number of interviews with planning, public works, and budgeting actors involved in the effort at assembling the infrastructure bond issue.

25. In my experience, I find technical and engineering analysis too coarse-grained to generate project rankings based on health and safety criteria. At best, it can cluster projects into broad groupings like a triage. When this is done, there generally are more "most

needy" projects than available resources. The choice of which "most needy" ones to prioritize is best left to policy or political criteria.

26. I am not denying that any need of their political sponsor would take a higher priority. Rather, this argument goes beyond that political interpretation and suggests that even with appropriate political controls, the bias would still exist.

27. To state it simply, small business owners represent a negligible vote, are modest political financial contributors, and generally opt out of the process of trying to influence politics.

28. Rivlin (1992, p. 267). Rivlin adds that "Mell and his allies opposed the proposal precisely because it would make Washington look good."

29. I accompanied the mayor on these walks. Generally, city staff preceded us, distributing an informational brochure describing costs and projects. Along one street, I was stopped by an older man who asked if rebuilding his street would really cost only $2.50 per year on his tax bill. This was our publicly estimated cost for a prototypical Chicago bungalow. I reassured the individual, grabbed his brochure, and did testimony by signing my guarantee. At that moment, the mayor walked up, saw what I was doing, and added his name to mine. A sheepish opposition alderman walking with the mayor added his signature.

About 2 years later, I received a note from the old man, accompanied by photocopies of our signatures and his tax bill. His taxes for the infrastructure program had gone up $4.00 per year. His note said "I knew it would cost more than you said, but I'm glad you did it anyway."

30. I was a party to the meeting.

31. By way of comparison, recall that the World's Fair Authority, at the height of its hyperbole, claimed a one-time impact of $2.5 billion to be derived from the Fair.

32. This section is adapted, with permission of the publisher, from Mier and Moe (1991, pp. 85-91).

33. I had leadership responsibility for this effort, hence culpability for the failure.

34. This belief was based on several failed attempts to tap this tax base for revenues to support social services.

35. The Chicago Central Area Committee is a long-standing civic association comprised of central area property owners complemented by members of the development and architectural communities. It has a tradition of strong advocacy for downtown growth, and has a reputation of being the "shadow" city planning department. It is identified by Shlay and Giloth (1987, p. 314) as one of the "core" groups in Chicago's growth coalition.

36. Again, I led this effort.

37. The group leading the opposition was the Interfaith Organizing Project (IOP). IOP subsequently represented the community in negotiations with the owners of the Chicago Bulls and Chicago Blackhawks to build an indoor arena at the same west-side site. See Mier (1991b, p. 15).

38. The community organized under the banner of "C.U.B.S."—"Citizens United for Baseball in the Sunshine".

39. It remains a mystery why the Archdiocese chose to become involved in this public controversy instead of scores of others revolving around issues of social justice and fairness. Nonetheless, the Cardinal's intervention, in a city with an extremely high population of Roman Catholics, carried enormous political weight.

40. With regard to historical narrative, White reveals enormous disciplinary pressures to narrate in a more scientific fashion. Such narration, he argues, is sanitized of emotion and imagination, the stuff of social action. White (1987, pp. 58-82, 185-213).

41. I have been inspired to this approach by Berger (1979, pp. 195-213) and Gordimer (1989, pp. 104-110).

42. Schön (1979, pp. 254-283) argues that generative metaphors are both frames of reference and processes for bringing new perspectives into existence. They are ways of seeing one thing as another, and in so doing, enabling the redefinition or resetting of a problem.

43. Marris (1987, p. 144) proposes "reproduction" as a metaphor that invites attention to patterns of relationships: "The reproduction of social relationships requires continual effort, attention and collaboration in which everyone, consciously or unconsciously, is involved every day of their lives."

Similarly, in a fascinating introduction to organizational analysis based on the use of metaphor, Morgan (1987, pp. 233-272, 371-376) introduces the metaphor of "flux and transformation" as connoting holistic group behavior in a context of a dialectic tension between organizational maintenance and change.

44. Schön (1983) says this an essential characteristic of a reflective professional practice.

REFERENCES

Alkalimat, A., & Gills, D. (1989). *Harold Washington and the crisis of black power.* Chicago: Twenty-First Century Books and Publications.

Berger, J. (1979). Historical afterword. *Pig earth.* New York: Pantheon.

Brehm, R. (1991). The city and the neighborhoods: Was it really a two-way street? In P. Clavel & W. Wiewel (Eds.), *Harold Washington and the neighborhoods: Progressive city government in Chicago, 1983-1987* (pp. 238-269). New Brunswick, NJ: Rutgers University Press.

Chicago Central Area Committee. (1973). *Chicago 21: A plan for Chicago's central area communities.* Chicago: Author.

City of Chicago. (1982). *Chicago 1992: Comprehensive plan, goals and policies, and ten year capital development strategies.* Chicago: Author.

City of Chicago. (1984). *Chicago works together: The 1984 development plan.* Chicago: City of Chicago, Department of Economic Development.

Clavel, P. (1986). *The progressive city: Planning and participation, 1969-1984.* New Brunswick, NJ: Rutgers University Press.

Clavel, P., & Wiewel, W. (Eds.). (1991). *Harold Washington and the neighborhoods: Progressive city government in Chicago, 1983-1987.* New Brunswick, NJ: Rutgers University Press.

Committee to Elect Harold Washington. (1983). *The Washington papers.* Chicago: Author.

Community Workshop on Economic Development (CWED). (1982). *Draft platform.* Chicago: Author.

Corral, L. M. (1992). Assessing the state of the infrastructure. *Chicago Enterprise, 6*(11), 14-19.

Friedmann, J., & Weaver, C. (1979). *Territory and function: The evolution of regional planning.* Berkeley: University of California Press.

Gills, D. (1991). Chicago politics and community development: A social movement perspective. In P. Clavel & W. Wiewel (Eds.), *Harold Washington and the neighborhoods: Progressive city government in Chicago, 1983-1987* (pp. 34-63). New Brunswick, NJ: Rutgers University Press.

Goldsmith, W. W., & Blakely, E. (1992). *Separate societies: Poverty and inequities in U.S. cities.* Philadelphia: Temple University Press.

Gordimer, N. (1989). A writer's freedom. *The essential gesture: Writing, politics & place.* London: Penguin.

Green, P. M., & Holli, M. G. (Eds.). (1987). *The mayors: The Chicago political tradition.* Carbondale: Southern Illinois University Press.

Greer, J. L. (1984). *Capital investments and the 1992 Chicago world's fair.* Chicago: Chicago 1992 Committee.

Hollander, E. (1991). The department of planning under Harold Washington. In P. Clavel & W. Wiewel (Eds.), *Harold Washington and the neighborhoods: Progressive city government in Chicago, 1983-1987* (pp. 121-145). New Brunswick, NJ: Rutgers University Press.

Holli, M. G., & Green, P. M. (1989). *Bashing Chicago traditions: Harold Washington's last campaign.* Grand Rapids, MI: William G. Eerdmans.

Judd, D., & Parkinson, M. (1990). Patterns of leadership. In D. Judd & M. Parkinson (Eds.), *Leadership and urban regeneration: Cities in North America and Europe* (pp. 295-307). Newbury Park, CA: Sage.

Krumholz, N. (1991). Equity and local economic development. *Economic Development Quarterly, 3*(4), 291-300.

Krumholz, N., Kostigan, P., & Keating, D. (1985). A review of Chicago works together: 1984 development plan. *Journal of the American Planning Association, 51*(3), 395-396.

Lakoff, G., & Johnson, M. (1980). *Metaphor we live by.* Chicago: University of Chicago Press.

Luke, J. S., Ventriss, C., Reed, B. J., & Reed, C. M. (1987). *Managing economic development: A guide to state and local leadership strategies.* San Francisco: Jossey-Bass.

Marris, P. (1987). *Meaning and action: Community planning and conceptions of change.* London: Routledge & Kegan Paul.

McClory, R. (1986). *The fall of the fair: Communities struggle for fairness.* Chicago: Chicago 1992 Committee.

McCullom, R. (1992). *A new "west side story".* Chicago: Interfaith Organizing Project.

Mier, R. (1989). Neighborhood and region: An experiential basis for understanding. *Economic Development Quarterly, 3*(2), 169-174.

Mier, R. (1991a). Baseball brinkmanship: Any economic development impact. *Michigan Partnership for Economic Development Assistance News, 4*(3), 1.

Mier, R. (1991b, July 15). A stadium deal that won't leave residents in the cold. *Crain's Chicago Business,* p. 15.

Mier, R., & Fitzgerald. J. (1991). Managing local development. *Economic Development Quarterly, 3*(3), 268-279.

Mier, R., Fitzgerald, J., & Randolph, L. A. (1993). African-American elected officials and the future of progressive political movements. In D. Fasenfest (Ed.), *Economic development policy formation: Experiences in the United States and the United Kingdom* (pp. 90-108). New York: St. Martin's Press.

Mier, R., with Giloth, R. P., Moe, K. J., et al. (1993). *Social justice and local development policy*. Newbury Park, CA: Sage.

Mier, R., & Moe, K. J. (1991). Implementing strategic planning: Roadblocks to reform, economic development and equity. In P. Clavel & W. Wiewel (Eds.), *Harold Washington and the neighborhoods: Progressive city government in Chicago, 1983-1987* (pp. 64-99). New Brunswick, NJ: Rutgers University Press.

Mier, R., Moe, K. J., & Sherr, I. (1986). Strategic planning and the pursuit of reform, economic development, and equity. *Journal of the American Planning Association, 51*(3), 299-309.

Mier, R., Wiewel, W., & Alpern, L. (1992). Decentralization of policy making under mayor Harold Washington. In K. Wong & L. Lynn (Eds.), *Policy innovation in metropolitan Chicago*. Greenwich, CT: JAI Press.

Morgan, G. (1987). *Images of organization*. Newbury Park, CA: Sage.

Reardon, K. (1990). *Local economic development in Chicago, 1983-1987: The reform efforts of Mayor Harold Washington*. Unpublished doctoral dissertation, Cornell University.

Rivlin, G. (1992). *Fire on the prairie: Chicago's Harold Washington and the politics of race*. New York: Henry Holt.

Schön, D. (1979). Generative metaphor: A perspective on problem-setting in social policy. In A. Ortony (Ed.), *Metaphor and thought* (pp. 254-283). New York: Cambridge University Press.

Schön, D. (1983). *The reflective practitioner*. New York: Basic Books.

Shlay, A. B., & Giloth, R. P. (1987). The social organization of a land-based elite: The case of the failed Chicago 1992 World's Fair. *Journal of Urban Affairs, 9*(4), 305-324.

Squires, G. G., Bennett, L., McCourt, K., & Nyden, D. (1987). *Race, class, and the response to urban decline*. Philadelphia: Temple University Press.

Warner, S. B. (1972). *The urban wilderness*. New York: Harper & Row.

We start planning today: City gets go-ahead for 1992 Fair. (1982, June 24). *Chicago Tribune*, p. 1.

White, H. (1987). *The content of the form: Narrative discourse in historical representation*. Baltimore, MD: Johns Hopkins University Press.

Wiewel, W. (1983). *Employment and the 1992 World's Fair*. Chicago: Chicago 1992 Committee.

Wiewel, W., & Clavel, P. (1991). Conclusion. In P. Clavel & W. Wiewel (Eds.), *Harold Washington and the neighborhoods: Progressive city government in Chicago, 1983-1987* (pp. 270-293). New Brunswick, NJ: Rutgers University Press.

World's Fair Advisory Committee. (1983). *A report to Mayor Harold Washington*. Chicago: City of Chicago.

4

Conditions of Confusion and Conflict: Rethinking the Infrastructure-Economic Development Linkage

CLAIRE L. FELBINGER

From the beginning public infrastructure has been linked with economic development. But, by the middle of the nineteenth century, as the causes of disease became better understood, the role of government—state and local government especially—in protecting the health and safety of citizens against the excesses of the market increased, changing the purposive direction of public infrastructure policy at all levels. The result of this is a decidedly uneven and confusing and often conflictual public infrastructure policy.

■ Concepts and Current Linkages

To begin, the concepts of infrastructure and economic development need to be defined, their changing roles summarized, and their current linkage specified.

Infrastructure as Public Works

More than 200 years ago, Adam Smith suggested that the functions of government should be limited to providing for the national defense, affording legal protection, and undertaking "indispensable" public works (Smith, 1976).[1] These indispensable public works, according to Smith, include those that facilitate commerce of the society and education, and that support the national defense. In terms of commerce, these include roads and navigable canals. These systems made commerce between trading companies and the forts and garrisons more convenient while also supporting the national defense.

The infrastructure system itself has come to include a subset of public works: described generally as "the permanent physical installations and facilities supporting socioeconomic activities in a community, region, or nation" (Hite, 1989). These physical systems, or public works, are usually identified as those physical structures and processes that support community development and enhance physical health and safety—roads, streets, bridges, water treatment and distribution systems, wastewater treatment facilities, irrigation systems, waterways, airports, and mass transit—as well as other utility services that are regulated by the public sector. They are, arguably, the most important and visible public works systems. They are distinguished from private-sector infrastructure such as manufacturing facilities. Hence, public infrastructure traditionally includes *public, physical systems* that ensure a healthy quality of life and enhance movement of people, goods, and services.

Infrastructure as Maintenance and Finance

John Eberhard and Abram Bernstein conceptually expand the physical nature of infrastructure by suggesting that each infrastructure system has three generic components:

> [A]n *origin*, where some desired material or service is created or generated; a *destination*, where the material or service is transformed into usable products or converted into waste products for disposal; and a *distribution network* for transporting things from where they are generated to where they will be processed. (Eberhard & Bernstein, 1985)

Therefore, infrastructure systems must be *designed, built, connected, delivered*, and *consumed*. They must also be *maintained* in order to preserve the structural and service quality integrity. Americans have done a good job designing and building infrastructure systems. Virtually all urban citizens are connected to these systems. Moreover, because the historical evolution of infrastructure systems (outlined below) has rendered access to these systems an absolute right of citizenship, they tend to be priced below market values (as in water and sewer services), or their price is highly regulated (as in natural gas and electricity).

It is the lack of commitment to maintenance that contributes most to the crumbling urban infrastructure. There are a number of reasons for this neglect. First, the pricing is often too low to pay for full maintenance. Crumbling structures do not occur overnight; they are a function of years of financing neglect. Second, many infrastructure systems are underground, hidden from view. As long as they are working, they are assumed in good condition. Until very recently, with the advent of new technologies (e.g., video recorders in sewer pipelines) and nondestructive testing, it has been difficult to monitor the integrity of those systems.

Third, even if maintenance were planned, the combination of fiscal stress and federal funding standards is often enough to ensure practices of deferred maintenance. Putting off expenditures until systems fail is like rolling dice—you take your chances that the facility will hold out and that no citizen is injured.

New York City provides many examples of the impact of deferred maintenance and increased risk. Recently, the City has been engaged in one of the world's largest capital projects, the construction of a new water tunnel from the Catskill Mountains. This project was undertaken because engineers could not test existing valves on the two original lines. The lines were in such a state of disrepair that engineers were afraid that once they closed the valves, they would not be able to open them again. They replicated a system that earlier preventative maintenance could have possibly saved. They decided it was not worth the possibility of losing one of the existing lines, so they built another.

In the city itself, water mains pose another problem. At the end of the Koch administration, there was an articulated policy in place concerning the replacement of water mains. To save money, the mains would be replaced only when they broke—replacement by emergency. However, the old mains were packed in asbestos, so waiting for them to break resulted in explosions that sent asbestos into the environment. The potential liability from such an environmental threat was greater than the cost of replacing the mains in a more orderly manner. In spite of this, fiscally strapped politicians were willing to take their chances on deferred maintenance.

Politicians also bow to pressure from citizens to take risks. City engineers closed the Williamsburg Bridge in New York City, citing that the structure could no longer sustain existing traffic. Pressure from commuters forced politicians to reopen the bridge. Not until the press demonstrated that many of the structure's I-beams were so corroded that they were the thickness of a quarter was the bridge closed again.

Until very recently, national infrastructure policy has added to the deemphasis on maintenance by supporting capital financing plans for the construction of new facilities and reconstruction of failed systems at the expense of maintaining existing ones.[2] Local managers used free federal dollars to build as many facilities and structures as there were federal grants available. For local leaders, construction of new projects is viewed as testimony to their political success, but routine maintenance is considered just that—routine—with little political payoff. Moreover, in times of fiscal austerity, federal programs made it fiscally prudent for a manager to defer maintenance until systems became eligible for federal reconstruction grants—when the systems failed. Even though fewer dollars would have been needed to properly maintain the old system, the availability of federal dollars for new construction made managerial decisions to apply local funds to maintenance unrealistic.

A classic example of how well-meaning federal policies result in poor management practices and decisions is the implementation of the Airport and Airway Trust Fund. These funds can be used only for substantial rehabilitation or replacement of aviation facilities. Repairing these facilities is not a fundable activity. As a result, airport managers may decide against repairing runways, letting them deteriorate to the extent that they need more costly rehabilitation. Why would a good manager wait? Does waiting not lead to a waste of taxpayers' dollars in such a case? Because managers of local facilities are responsible to local taxpayers, it is prudent to leverage local dollars with federal funds. Therefore, if a manager repaired the runways, 100% of the expenditures would have to come from locally generated funds. There is no question that it is a better managerial practice to repair the runways, resulting in longer life spans and longer intervals between rehabilitation episodes. However, it may be a better local fiscal strategy to apply for reconstruction funds—100% federal funds—and let the federal government pay.

Infrastructure as Administration and Management

Eberhard and Bernstein (1985) also suggest that the distribution of infrastructure is provided not only by pipes, pumps, roads, and bridges, but also by administrative and managerial networks that plan, build, maintain, and finance the physical systems. Thus, the definition of infrastructure is expanded to include the *human infrastructure* necessary to make these systems work.

Historically, building and managing the network of infrastructure has been almost exclusively the purview of civil engineers. By virtue of their training, civil engineers quite effectively perform a portion of the functions that Eberhard and Bernstein suggest. They design, build, connect, and deliver infrastructure systems and services. However, according to John Peterson (1991), "the existing ways of designing, selecting and paying for public works leads to inefficiencies." By this he meant that those who continue to do things as they have "always been done" seldom take advantage of efficiencies from use of new materials, processes, or creative financing opportunities. Building, designing, and paying for infrastructure does not make the system complete. There is little attention paid to the human infrastructure function of public works service delivery by many who oversee infrastructure systems. The dominance of traditional civil engineers in infrastructure, consequently, has contributed to the deterioration of systems and the lack of attention to infrastructure issues.

Specifically, there are several aspects to this problem. First, civil engineers are trained to build. Nowhere in their formal education are they trained to maintain. Therefore, all emphasis is placed on the design and construction of new structures. Second, they are not schooled to think about the health, safety, and convenience impact that comes as a result of these built systems. Rather, they are trained in "structures," "hydraulics," "geotechnics." The social impacts of their work are never addressed. Third, civil engineering schools have failed to provide any grounding in the larger picture of managing these networks, including budgeting, finance, human resources, communication, and maintenance. It is not that they have no regard for management of systems, or for the human infrastructure charged with managing infrastructure, they just never think about it—it is not in the equation. As a result of this, and compounded with the political and fiscal problems mentioned previously, infrastructure is often left, if not in ruins, at least in a state of confusion and deterioration.

The quality of infrastructure rests not only on the quality of the design and construction of systems, but also on the quality of the human infrastructure that supports and manages it. Infrastructure, then, consists of the indispensable public *physical* and *human* systems that provide the nation with services such as potable water, wastewater treatment, transportation systems, and utility services. The argument in this chapter is that these systems are the most essential, the primal foundations that support modern civilization. As such, all citizens have a right to expect a high quality and sustained integrity of these systems. Oliver Byrum (1991) argues that low-quality infrastructure services should be unacceptable in a democratic society. Everything else, includ-

ing economic development, therefore, should be a secondary consideration until the essential systems are available to all.

Economic Development

The American Economic Development Council (AEDC) defines
economic development as "the process of creating wealth through the
mobilization of human, financial, capital, physical, and natural resources to generate marketable goods and services" (AEDC, 1984).
Richard Bingham, Edward Hill, and Sammis White put it more simply:
"[I]t is the creation of jobs and wealth" (Bingham, Hill, & White, 1990).
The role of economic development is building the private city by
*re*developing central cities or developing new areas and industries. It is
the accumulation of wealth for the private economy.

Even though the AEDC was organized in 1926, Bingham, Hill, and
White contend that economic development did not become important to
Americans until the recession of 1981 and the increase in foreign
competition that blossomed in the mid-1980s (Bingham et al., 1990).
Economic development became the prevailing policy focus of a vast
majority of elected and appointed officials at all levels of government.
Economic development denoted growth; the desire for growth created
a consensus across a wide range of elites. This was (and is) the case
even when the local elites historically disagreed on other issues (Logan
& Molotch, 1987).

Economic development (or "health") of a region can be assessed
using a variety of longstanding theories or approaches, such as neoclas-
sical, economic base, location, central place, attraction, agglomeration,
and product life cycle theories (Blakley, 1989). Depending on the
theoretic bent of regional officials, certain economic development strategies are more attractive than others. As early as 1981 the urban policy
question was not, "Do we want economic development?" but "How do
we attain it and what policies should the government propose to enhance
it?" Consequently, economic development became the central feature
of the federal Urban Development Action Grant Program, along with
debates over Enterprise Zones. "Job retention" became a major state and
local issue and "industrial attraction" strategies blossomed (Birch,
1987; Bluestone & Harrison, 1982; Etzioni, 1983; Jones & Bachelor,
1986; Markusen, Hall, & Glasmeier, 1986; Reich, 1983). Economic
development had a following among politicians, workers, and a small
though influential group of academics.[3]

Current Linkages

Like latter-day Adam Smiths, scholars and politicians moved to make "indispensable" the link between infrastructure and economic development: New infrastructure investment would be justified as an economic development effort. Here the role of public infrastructure has shifted from one that serves people (health, safety, and convenience) to one that primarily serves private interests (jobs and wealth). Although not ignoring the obvious importance of public infrastructure to economic development, this chapter will critically assess the impact of this change in role as it explores the historical evolution of the role public infrastructure has played in American society. I will assess the current empirical efforts to link infrastructure investment and economic development. I will also argue that the primacy of economic development as a public policy increases the risks to the health, safety, and convenience of certain identifiable individuals and, ultimately, to collective health and safety through the destruction of the environment.

■ Public Infrastructure in Historical Perspective

This section demonstrates that the role of public infrastructure has evolved over time. Initially, infrastructure aided mercantile efforts— early economic development. Health and safety connections with infrastructure and public works emerged only with advances in our understanding of the transmission of disease. Legal decisions and moral thinking reinforced the primal linkage with a healthy city and a clean environment. However, laissez-faire governmental policy allowed infrastructure investment to be market driven once again. Economics and health and safety were then both linked to the investment in infrastructure during the New Deal. Current policy priorities have all but forgotten the health and safety aspects of infrastructure investment.

Frontier Cities:
The Emergence of Health and Safety in the 1800s

American citizens have always depended on public infrastructure for their survival and prosperity. The early colonists relied on harbors to maintain contact with their homeland. Later, with westward expansion, transportation networks were required to move people and products across the broad expanse of the country. Public infrastructure played a major role in the urban/industrial development of the nation. Initially, the role was almost wholly mercantile—serving the interests of the mercantile class to move its products to market.

During the 1800s industrialized cities were not visions of beauty—their streets were truly the repositories for the garbage of the mercantile city. The very first paved streets, constructed of boards, were introduced primarily to accommodate the transport of goods over muddy and littered streets. Secondarily, they also allowed people to walk in the city without stepping in rotting refuse.

The lack of sanitation also affected ground and surface water. As cities industrialized, the incidence of pollution and disease grew accordingly. The very first water improvements, unfortunately, had little to do with promoting health and safety. The linkage of polluted water to the incidence of disease was unclear. The first water systems, for example, were used to improve travel in the city; to flush garbage out of the way of commercial and business transport. Second, controlling the flow of water helped in the suppression of fires. In fact, city officials of the 1800s were not concerned with distributing water to citizens directly. Water systems were viewed as part of the economic infrastructure of the city.

But the horrendous sanitary conditions inspired urban utopian authors to characterize the great cities of the future as ones in which threats to the health and safety of citizens would be eradicated. These "urbtopians," as Stanley Schultz (1989) described them,[4] portrayed the city of the future as one in which the prevailing threats to health and safety of citizens (including massive outbreaks of cholera and typhoid) would be eradicated by technological improvements coincident with public infrastructure improvements. The urban public works of the future were described as integrated systems of sewers, air and automobile transportation, heating and air conditioning, and rapid communications. The urban future was one characterized by sanitary conditions for all citizens. Technology would clean the water, remove sewage, and dispose of solid wastes.

The Era of Health and Safety:
Moral Issues

If the urbtopians made the futuristic linkage between infrastructure and health and safety, antebellum urban planners made a moral linkage. In an era of rapidly changing notions on how diseases were transmitted, urban environmentalists were most concerned with five conditions: "epidemics; the unsavory condition of urban streets; the ways in which cities disposed of garbage, wastewater and animal wastes; increasing air pollution; sudden appearance of slum housing" (Schultz, 1989).

These reformers were convinced that dealing with these maladies would not only make citizens healthier, but would also make them more "moral": A clean environment led to "good people" (Schultz, 1989). Such people would locate in places that were free of disease—places where they would be healthier and more productive. To invest in infrastructure meant the promotion of the health and safety and morals of the citizens. This movement, led by sanitarians, engineers, urban planners, and the growing urban middle class, equated the efficiency of infrastructure systems with the quality of the entire civilization.[5] Such lofty goals for public infrastructure underscored its key role in ensuring health and safety. For example, in Chicago, Ellis S. Cheesebrough, "Father of Chicago's Sewer System," was entrusted in 1855 to "improve and preserve" the citizens' health (Griffith, 1974). Conversely, cities that did not heed the call for infrastructure improvements, like Baltimore, continued to suffer higher death and disease rates than those that instituted infrastructure improvements. It was becoming clear that the introduction of potable water systems dramatically decreased the incidence of the waterborne diseases of cholera and typhoid. Increased public works meant increased quality of life.

This is not to say that such technological reforms were not supported by those promoting cities as "containers of business activity" (Schultz, 1989). Clean cities were ones in which there were no quarantines, and this was good for business. At the same time, business interests were keenly aware that the moral environmentalists blamed them in large part for the poor conditions in the city, from slum housing to pollution. Economic development was affected by infrastructure investment in this era; however, it did not necessarily drive it. Indeed, rapid economic change was more often than not viewed as a key source of urban health deterioration and safety concerns, not a solution to the oppressive conditions of the urban environment.

The Era of Health and Safety: Legal Issues

Eminent Domain and Police Powers. It was the legal system that established a formal connection between public works and protecting health and safety. The concept of eminent domain, practiced as early as 1820, was adjudicated by Justice Lemule Shaw in 1834. In the Wellington case (*Wellington et al. Petitioners*, 1834) the petitioners wished to lay a highway over lands designated by the state legislature as public lands, the activities on which were regulated. Although he did not allow the highway development due to state sovereignty, Justice Shaw began the crafting of eminent domain as it refers to seizure of private lands for public uses:

> The subject of highways, bridges, canals, and other public works, for facilitating communication between one part of the territory of a State and another, are peculiarly within the province of the sovereign power of the State, represented and exercised by the legislature. In cases of more frequent recurrence, as those of roads and ferries, it has been the practice of the legislature of this Commonwealth to provide for the establishment of this common class of public easements, by delegating the power to some properly constituted body. But in the more rare and often the more important and interesting cases, as those of bridges over navigable waters, canals, and more recently, railroads, the legislature has exercised the power in each particular case, and also in that large class of most important highways, turnpikes, the expense of which is borne by individuals in the first instance and repaid by tolls. There seems to be no objection in principle to the exercise of this authority on the part of the legislature. (*Wellington et al.*, 1834)

Justice Shaw and the court viewed government's protection of health and safety through public works as a proactive foundation in spite of the nuisance laws. In *Edmund Baker v. City of Boston* this health and safety perspective was combined with police powers to ensure public health over private interests. The case involved Boston's filling in of a navigable canal and holding that a slaughterhouse could not continue to pump wastes (a nuisance) in the area. The court held that

> It is not only the right but the duty of city government of Boston, so far as they may be able, to remove every nuisance which may endanger the health of the citizens.
>
> And they have the necessary power of deciding in which manner this shall be done, and their decision is conclusive, unless they transcend the powers conferred on them by the city charter.
>
> Police regulations to direct the use of private property so as to prevent its being pernicious to the citizens at large, are not void although they may

in some measure interfere with private rights without providing for compensation.

The property of a private individual may be appropriated to public use in connection with measures of municipal regulation, and in such case compensation must be provided for, or the appropriation will be unconstitutional and void.[6]

Other cases involving police powers would be justified on the same basis.[7]

According to Schultz, in the 1800s, to "police" the city:

[M]eant to prevent actions that threatened public order, safety, health, or convenience, and to promote the means of securing them. . . . For the sake of public health standards, street improvements, the provision of water and sewer systems, the recruitment of professional police and fire departments, and dozen of other amenities, a growing number of urbanites championed a stronger than traditional involvement of local government in daily life. (Schultz, 1989)

Hence, governmental responsibility expanded to include proactive policies designed to ensure the health and safety of the citizenry.

The prominence of health and safety over economic development is no more clearly specified than in the landmark U.S. Supreme Court case, *Village of Euclid v. Amber Realty Company* (1926). In Euclid a private developer wanted to extend the "natural development" of industries from the east side of Cleveland to its eastern suburbs. The Village of Euclid did not want this industrial expansion that, incidentally, would clutter its lakefront. The village argued that the expansion would pollute the community, adversely affecting the health and safety of its residents. The realty company argued that the city's land use planning and zoning decision was not based on health and safety but on aesthetics, discriminating against its right to profit through industrial development. The Supreme Court ruled in favor of the village, citing primacy of health and safety over economic development.

Thus, where the courts have been a place of recurring conflict between policies preserving the health, safety, and convenience of citizens and those pursuing private economic goals, they tended, quite often and quite clearly, to resolve competing interests in favor of the position of health and safety (as in the Shaw ruling). Municipal corporations have long been encouraged by the courts to regulate land use in a way consistent with the health and safety needs of the people: the building and operations of water and sewer systems, institution of building codes, and provision for urban lighting, for example, as legitimate features of such policy.[8]

The End of Laissez-Faire Public Works Policy[9]

Regardless of the courts' insistence that it was the responsibility of local governments to provide infrastructure systems to enhance the health and safety of citizens, public action on these items was usually prompted by the needs of private-sector decision makers. By the end of the nineteenth century rapid urban growth was the product of rapid changes in investment in industrial infrastructure. People were drawn to the cities as the nation's urban population increased from 19.8% in 1860 to 45.7% in 1910 (Bingham, 1986). By the late 1920s the new urban environment teetered unstably on the rocky foundations of increasingly fragile investments in private infrastructure. The result was a laissez-faire approach to public works that was just as uneven in its health and safety standards as it was in the marketplace.

All of this came to a halt with the Stock Market crash of 1929. Unemployment soared from 3% in 1929 to nearly 25% in 1933. President Herbert Hoover, in an action that was too little too late, approved the 1932 Relief and Construction Act, which allocated $300 million for public construction projects designed to get the economy back in line. Investment in public works had become a source of economic recovery. No longer were public works viewed as sources of either health or development—their construction was to be part of a national plan of economic redistribution and recovery. President Franklin D. Roosevelt added a variety of agencies, whose major activities involved public works and infrastructure projects: Civilian Conservation Corps, Tennessee Valley Authority, Federal Emergency Relief Administration, the Civil Works Agency—and the largest two—Works Progress Administration (WPA) and the Public Works Administration (PWA).

More than 75% of all WPA funds were spent on construction projects. This included either building or improving nearly 500,000 miles of roads, erecting 78,000 bridges, and constructing thousands of schools, hospitals, courthouses, and sanitation facilities. These activities employed about 8.5 million people. The PWA built federal facilities and made loans and grants for large, durable facilities. All but 2 of the nation's 3,071 counties benefitted from PWA: 70% of the schools in the country were built over a 7-year period from 1933-1939 and 65% of all courthouses were built during the same period (Armstrong, 1976).

Infrastructure programs were used to revitalize sagging economies even after World War II. The Area Redevelopment Administration (ARA) was established in 1961, in the Commerce Department, to provide technical assistance and loans for local public projects (particularly, though not exclusively, for those that would attract private development). The Appalachian Regional Development Act, which began in 1965, was a product of the Senate Public Works Committee and still acts as a regional economic development agency. In 1965 the Public Works and Economic Development Act replaced ARA with the Economic Development Administration (EDA), funding public works projects in depressed areas. The water supply systems, sewer lines, and roads that serve business and industry in thousands of industrial parks across the country have been constructed using EDA funds.

Public works and infrastructure projects served social as well as economic ends. There were antidiscrimination clauses, accompanied by compliance provisions in national public works contracts, as early as 1941 (defense industry contracts). The Fair Employment Practices Committee was in place in 1942. Contract compliance continued under committees established in 1951 and 1953. Public works contracts were also covered under the Equal Employment Opportunity Act of 1961 and expanded with the 1964 Civil Rights Act. Finally, the Davis-Bacon (prevailing wage) Act assured employees of a fair wage when working on projects funded with federal money.[10]

Consequently, from the 1930s through the 1970s, the federal government's approach to public infrastructure was one that targeted economic decline while it incrementally assisted in responding to the growing social agenda of the city as well. The use of public infrastructure as a tool of economic recovery was solidified during this period of increasing federal intervention, setting the stage for the contemporary linkage of public infrastructure with economic development.

■ Infrastructure Investment Today: Building the Private City at Whose Expense?

From the beginning, public infrastructure has been linked with economic development, but, as the causes of disease became better understood, the role of government in protecting the health and safety of citizens against the excesses of the market changed the purposive direction of public infrastructure policy as well. The courts made it the government's responsibility to ensure health and safety of citizens, even if this meant that the government would have to acquire property through eminent domain, or regulate activity (even economic development activity) through police powers. In 1905 municipal reformer Frederic Howe suggested that these infrastructure systems were too important to be left to the wiles of the private sector because they had become a "necessity to the life, health, and comfort, convenience, and industry of the city" (Howe, 1967).

The perceived role of infrastructure investment had expanded, playing a big part in the development of our nation socially, healthwise, as well as economically. Although public works were perceived to provide the framework in which economic activity could take place, they were not undertaken as acts of capital accumulation themselves; they were not necessarily what we currently call economic development: that is, the creation or retention of real jobs and industries. All this has changed over the past few decades, reflecting

1. the decline and change of federal role in cities in general and in public infrastructure in particular;
2. the increase in state and local participation in public works; and
3. the decentralization of public intervention and decline of central cities.

The decline in federal funds and the concomitant increased dependency on state and local resources has led to a form of spatial and fiscal triage, where the projects most often built are those that can attract some form of funding and can be built in areas exhibiting the fastest economic growth. For instance, short-term jobs are created when an expressway interchange is constructed for a facility that will produce real jobs and real wealth. Such infrastructure investment is justified because it assures a real economic development payoff. At the same time maintenance of existing structures and systems in city neighborhoods offer no revenue stream or economic payoff, and in these perilous fiscal times there is little incentive for the fiscally overburdened city to fund them. In this era of economic development, such maintenance is translated into "quality of life"—enhancing the "look" of a city—in the eyes of incoming business leaders. Investment in public capital occurs in the productive spaces of the private city—serving the office towers, sports complexes, and convention centers of the central business districts, and the new suburban satellites and neighborhoods of the urban fringe.

Health and Safety Concerns Today

But the issues of health and safety are far more than development rhetoric about the quality of life. Issues of health and safety are central to the decline and demands of current public works in all areas of the country—urban and rural. The decline of the physical condition of the built environment is also directly affecting the natural environment. Put another way, issues of health and safety, far from being strategically distributed in productive areas of economic development, are spread across the entire urban and rural region and can be translated into broad issues of the environment as well.

Urban Areas

The primary threat to health and safety in urban areas is the lack of maintenance and the overburdening of infrastructure systems. Take for example, the cases of wastewater treatment, drinking water distribution, and roads and bridges.

Wastewater. There are approximately 16,000 wastewater treatment facilities, which handle 37 billion gallons of sewage a day (Apogee Research, Inc., 1987). The vast majority of these serve citizens in urban areas. A Congressional Budget Office report indicates that gains in control of water pollution are a direct result of the expenditure of more than $90 billion in capital financing of wastewater treatment since 1972 (CBO, 1988). Much of this money was used to upgrade plants to secondary treatment. Eighty-two percent of the plants have met this standard. Capital financing from the federal government has dropped from $6 billion in 1980 to $2.4 billion in 1988 to $0 in 1991. State and local financing for operations and maintenance increased from $4.6 billion in 1980 to $6.8 billion in 1987 (U.S. Congress, Office of Technology Assessment [OTA], 1990). Still, the EPA's latest *Needs Survey* estimates a demand for $60.2 billion, or $250 per capita, to build enough secondary facilities to treat sewage for the current population (CBO, 1988). The EPA also estimates that it would cost an additional $84 billion in capital investments to bring all municipalities into compliance with the Clean Water Act (OTA, 1991).

It is not clear that urban governments have the fiscal capacity to keep up with these standards (through funding operations and maintenance) and still provide other needed urban services. This is especially the case as the federal government has ceased funding construction of wastewater facilities. If cities do make these wastewater investments, it is at the cost of services provided to neighborhoods. Richard Bingham has pointed out that during times of fiscal stress, some of the first services to be cut are those that receive no intergovernmental support, such as recreation, parks, libraries, cultural activities (and now construction of treatment plants)—services for neighborhood citizens (Bingham, 1986).

Local governments operate in a system of finite resources; they cannot roll over short-term debt. When resources are finite, any investment is zero sum to those activities that did not receive allocations. The alternative for supporting wastewater investment is to raise rates. Once again, this has a direct impact on urban dwellers. Moreover, because central cities have some of the oldest wastewater systems, the cost to repair and maintain existing facilities is highest. Because the economic payoff of investment in neighborhood sewer maintenance is not considered high (no new wealth or jobs), and because all investment is tied to economic development, no investment is justified.

Drinking Water. One-half of 1% of the nation's approximately 180,000 water systems provide potable water to 43% of the population (OTA, 1991). These, obviously, are urban water systems. Although water appears safe when tested at the treatment facilities, poor maintenance takes a toll on water system effectiveness and quality as the water moves toward its users. One study of eight urban water systems showed unaccounted rates (leakage) of from 1% to 37% (Mann & Beecher, 1989). The Office of Technology Assessment (OTA) reports that lining the aged pipes of a major city (maintenance) could have prevented a leakage rate of almost 40% of treated drinking water over several decades—more than enough to make up for shortages brought on by dry spells (OTA, 1991). This is not only costly, but harmful. With the exceedingly high continuation of leakage, the threat of contamination through infiltration of the "safe" treated water is also substantial—so much so that the OTA estimates that local governments will have to increase annual local capital outlays by 50% to comply with the standards of the Federal Safe Drinking Water Act. What makes this condition even more serious is the present practice of monitoring water quality that requires the municipality to only test the quality of the water at the treatment facility—there is as yet no requirement to test the quality of water at the tap—after it has traveled through the leaky pipes of the urban infrastructure (OTA, 1991).

Highways. The nation's highway system is composed of almost 3.9-million miles of roads and bridges, with an asset value of $480 billion (in 1984 dollars) (National Council on Public Works Improvement, 1988). A total of 690,947 miles are concentrated in urban areas (Pisarski, 1987). Total public expenditures for highways, streets, roads, and bridges have exhibited a steady decline since 1960, with an overall decline by 1984 of 9% below the 1971 peak.

In spite of that decline in investment, urban congestion is increasing. A study by the OTA found that across 15 major cities, only 1 (Detroit) experienced a decrease in congestion between 1982 and 1987 (OTA, 1989). The highest percentage change was in San Diego (38%), with an average percent change of 17.4%. In these cities the per capita cost of congestion ranged from $190 (Milwaukee) to $740 (Washington, D.C.). Delays due to highway congestion and poorly maintained roads in major urban areas already take an estimated toll of more than $30 billion annually—this is one half of what is spent nationally for all roads and bridges (OTA, 1991).

In turn this congestion also causes air quality problems, which affect the health and quality of life of urban residents. There are many environmental concerns linked with urban air pollution, including human health (increases in cancer, heart disease, lung disease); plant and animal health (increases in chronic injury); materials' and buildings' integrity (early aging and deterioration from sulphur dioxide in smog and ozone); aesthetic integrity (dimming visibility, obstructing views) (Henning & Mangun, 1989).

The condition of aging urban infrastructure systems, along with the preeminence of economic development as a policy ideology, affects the health, safety, and convenience of citizens in at least two ways. First, infrastructure investment is spatially targeted to areas and industrial interests central to a city's economic development—usually certain areas downtown and along the city's periphery (Byrum, 1991). In an era of scarce resources, the neighborhoods will be neglected, leaving them with poor streets, poor air quality, and questionable water quality (Byrum, 1991). Second, the presence of fiscal crises in many cities leaves calls for maintenance of some infrastructure systems (such as roads and bridges) to compete with upgrades mandated by state and federal governments (such as water and wastewater), and these, in turn, compete with other programs aimed at improving the quality of social and economic existence in the neighborhoods. The threat to urban health and safety can be real.

Once again, in the current intergovernmental environment cities are unable to shift their own source revenue from infrastructure investments funded by intergovernmental transfers to those not so funded. Revenue for that must be locally generated. Neither local agencies nor their constituents can choose to have effluent quality below those of federal standards; therefore, they must pay. There is no question of funds' fungibility or substitutability here because there is no infusion of outside funds (Wright, 1988). Necessarily, then, there is direct competition between mandated activities and nonmandated ones, because the fiscal environment is zero sum. And though this ensures quality wastewater treatment, it also ensures that other neighborhood expenditures will suffer.

When do neighborhoods "benefit" from infrastructure investments? According to John Logan and Harvey Molotch, it is when bureaucrats seek to site "devalued" infrastructure systems, such as sewage plants, jails, and halfway houses (Logan & Molotch, 1987). They give as an example the siting of solid waste incinerators and landfills in Houston. Although Houston is only 28% black (occupying the same percentage of the city's area), more than three fourths of the city-owned incinerators and all of its landfills are located in black neighborhoods. Harlem was transformed from an Irish enclave to a black ghetto by the dislocation of midtown-Manhattan blacks to make room for Pennsylvania Station. Logan and Molotch conclude that "poor people are double losers; they have the least to gain from the infrastructural development and much to lose by the choice of its location" (Logan & Molotch, 1987).

Rural Areas

Rural areas have inadequate infrastructure, poorly trained and limited numbers of professional staff, and few amenities to use for economic development. Also they receive even less in financial resources than urban areas for improving their physical systems. Rural officials struggle to provide adequate basic services with archaic infrastructure systems. For instance, rural water systems serve only 8% of the nation's population and produce 93% of all violations for contamination and

94% of all monitoring/reporting violations (National Council on Public Works Improvement, 1988). Of the 3.9 million miles of public roads that cross the country, 1.7 million miles remain unpaved (OTA, 1991), the lion's share of which are in rural areas.

Many rural communities still rely on individual septic systems to treat wastewater on site. These systems are effective and inexpensive when maintenance and soil conditions are good and space is available. However, rural systems are plagued with problems. One study suggests that 80% of rural counties reported system failure and potential ground and surface water contamination (OTA, 1991). The health and safety and convenience of rural citizens remain at risk in the current policy environment, which focuses most government resources on economic redevelopment downtown and new development at the city's "edge."

The Environment at Large

The primacy of economic development has contributed to the long-term physical decline and environmental degradation of both urban and rural areas. Highway systems have facilitated the move of the middle and upper classes from the city, and have stimulated new and expanding periods of congestion created by commuters traveling from their homes to downtown or peripheral places of employment. Transportation accounts for two thirds of all petroleum use—70% of which is used on highways (OTA, 1991). The intensive uses of fossil fuel technologies have been linked to the greenhouse effect and other damaging environmental impacts (OTA, 1991; also Conservation Foundation, 1982, pp. 73-74; Henning & Mangun, 1989, p. 221; National Research Council Board, 1983, pp. 22-34). Traffic volume and congestion contribute to air quality problems that further aggravate the decline of the city as a healthy and safe center for living. More than 76 million people live in counties that exceed federal ozone standards (Henning & Mangun, 1989). Wind and weather scatter these effects widely (Council on Environmental Quality, 1981). Here the notion of quality of life is more than an economic development nostrum—it is an essential feature of environmental balance, serving the health and safety of the entire community.

Those who have the most to lose, in the short run, are residents of inner-city neighborhoods and rural areas. First, they lack the resources with which to mobilize public policy in service of their needs. Second, the present practices of economic development policy at the local and state levels advantage industries and real estate development patterns that depend on the automobile. Further growth of an economy dependent on fossil fuel will continue to threaten the health and safety of rural and urban residents alike and will make it difficult to develop public infrastructure policies that place the environment and health and safety ahead of economic development. As Henning and Mangun note, governments tend to respond to these environmental problems in fragmented ways and that "[s]hort term considerations—economic development, progress, and growth—tend to determine policy" (Henning & Mangun, 1989).

■ The Empirical Practice of Linking Infrastructure Investment and Economic Development Measures

Historically, infrastructure investment has been linked to the development and economic redevelopment of the country. The latest emphasis on economic development is mirrored in economic analysis bent on empirically testing and reaffirming the linkage between public investment in physical capital and economic development. These studies have been used to justify arguments for increased public infrastructure expenditures.

Researchers who seek to estimate the strength of the linkage do so because they expect this type of investment to be different from other types of investments. Randall Eberts raises the linkage to a level of a theoretical given for the dual nature of the impacts of infrastructure investment: "First, public infrastructure provides the basic foundation for economic activity. Second, it generates positive spillovers; that is, its social benefits far exceed what any individual would be willing to pay for its services" (Eberts, 1990). Although the case for the social benefits, including health and safety, are made theoretically, they are seldom assessed in the literature because they are not included in conventional measures of output, such as gross state or national output. Two exceptions are William Nordhaus and James Tobin's "Measure of Economic Welfare" (Nordhaus & Tobin, 1972), and Robert Eisner's *Total Incomes System of Accounts* (Eisner, 1989).

In practice the economic development linkage has been examined at all levels of aggregation: national, regional, state, urban, and rural. One of the most widely cited works demonstrating the infrastructure-economic development linkage at the national level is a study by David Ashauer

(1989). He estimates that every $1 invested in public infrastructure nationally results in a $4 increase in the Gross National Product (GNP). Ashauer maintains that GNP is a measure of the nation's productivity and asserts that productivity slowdowns in the 1970s were a function of the decrease in public infrastructure investments nationally during that time. Ashauer's work has been widely used to argue for increases in public infrastructure allocations and expenditures, and widely criticized for being a gross overstatement. Though Munnell (1991) argues that the estimated impact should increase with each level of aggregation due to positive regional spillovers, which are not captured in more microanalyses, Eisner (1991) and Douglas Holtz-Eakin (1994), in independent analyses of Munnell's data, find that the best estimate of public investment's impact on private output or productivity is essentially zero—and in some cases negative. Holtz-Eakin goes so far as to dismiss such large positive effects as being "the artifact of an inappropriately restrictive econometric framework."

In the face of such criticism, other analysts are a bit less sanguine about the direct impact of public infrastructure on the economy: attesting to a concomitant downshift in certain key economic indicators at the same time there is a slowdown in spending for public works (Boskin, Robinson, & Huber, 1987; Costa, Ellson, & Martin, 1987). For instance, Kevin Duffy-Deno and Randall Eberts have found a small but positive and statistically significant relationship between shifts in private capital stock and per capita personal income in 28 metropolitan areas between 1980 and 1984 (Duffy-Deno & Eberts, 1989). They have determined that this linkage comes through actual construction of public capital stock, and through capital stock as an unpaid factor in the production process. Presumably, this unpaid factor includes concern for health and safety—the social spillovers. However, this is never explicitly revealed. They conclude that public-sector investment affects economic development both by increasing productivity and by reducing production costs. Similarly, Jay Helms (1985) has found that governmental expenditures on highways can have a positive effect on state personal income.[11]

Measures of public-sector infrastructure investment have also been linked to private-sector investments—as leverage for private investment. Kevin Deno has found that a 1% increase in public investment is associated with a .01% short-term and a 0.2% long-term increase in private investments (Deno, 1986). These findings were corroborated by Eberts and Fogarty, who found the relationships strongest in the older cities of the Midwestern region, that is, those which experienced much of their growth before the 1950s. These are the very cities that engaged in the rash of downtown redevelopment efforts of the 1970s and 1980s. There is no indication that these public-private sector partnerships extended to urban neighborhoods. However, Eisner argues that some negative results indicate that the infusion of public funds and labor is substitutive rather than additive or leverage producing (Eisner, 1991).

Infrastructure investment has also been shown to have a positive impact on manufacturing output and locational decisions. Jose da Silva Costa, Richard Ellson, and Randolph Martin report that public capital stock makes a statistically significant contribution to manufacturing output (Costa et al., 1987). William Fox and Matthew Murry (1989), Timothy Bartik (1985), and Randall Eberts (1990) all found strong, statistically significant, positive associations between the presence of a highway system, miles of road, and public infrastructure investment and firm openings and locational choices. These findings were more pronounced in urban/developed regions than in rural regions, as further demonstrated by Koichi Mera (1975) and predicted by Niles Hanson (1965). Of course, the building of these roads and the location of new business on the edge of urban areas only exacerbate environmental problems.

What Mera did find (and Hanson predicted) in the rural areas was a direct relationship between the introduction of new technologies and processes and *rural economic development*. The impact of direct capital investment on rural economies was far less dramatic. Technology transfer to rural areas had the effect of increasing economic productivity and reducing government spending. This finding appears to be consistent with the fact that rural service delivery systems are plagued by inadequate professionals as well as limited funds and poorly maintained facilities.

Where the empirical linkages have been made in rural areas, they have focused on targeted, nonessential systems (U.S. Department of Agriculture [USDA], 1990). By nonessential is meant those systems not essential for health and safety of citizens. For instance, Parker (1990) concluded that improving *telecommunications* should improve rural businesses' productivity (consistent with Mera and Hanson). Moreover, information-dependent industries may relocate in lower cost rural areas that already have appropriate telecommunications infrastructure in place. Once again, health and safety are not factored into the analysis. Perhaps this is because maintenance and operation of basic infrastructure systems are assumed to be adequate nationwide, despite evidence to the contrary. For example, many rural areas are in desperate need of potable water, because their groundwater is polluted by inadequate wastewater treatment (septic systems).

Other studies have verified the uneven linkage between infrastructure and economic development in rural areas. However, the focus, once again, is on economic indicators and not the social spillovers. For example, CONSAD Research assessed the linkage between public infrastructure investment in rural areas and found that 30% of the gain in real income in those areas between 1963 and 1966 was accounted for by investments in highways, waterway docks, vocational schools, and recreational facilities (CONSAD Research Corporation, 1969). On other rural investments, however, the picture is not so clear. Thomas Rowley, Neil Grigg, and Clifford Rossi indicate that investment in wastewater treatment *sometimes* results in positive economic development payoffs (Rowley, Grigg, & Rossi, 1990) (although it has an immediate positive effect on health and safety), though David Forkenbrock et al. find no strong evidence to suggest a linkage between transportation infrastructure and rural economic development (outside of airport investments) (Forkenbrock, Pogue, Finnegan, & Foster, 1990).

Despite the unevenness of the empirical linkages in the rural communities, David Sears, Thomas Rowley, and Norman Reid make what they term some "plausible justifications" for rural infrastructure investment outside of economic development:

> Many types of infrastructure, such as water treatment systems, hospitals, and fire stations, are needed to protect and promote the health and safety of rural residents. Without such key infrastructure investments, most of which are within the public sector, public health and safety could be seriously jeopardized.
>
> Many other facilities support services that make life more convenient and pleasant. These include parks and recreation facilities, libraries, museums, and other cultural facilities, especially in larger communities.
>
> The provision of services that enhance living conditions in a community—by promoting health and safety or by making life more convenient—may indirectly improve economic development prospects of the locality by increasing the area's attractiveness as a place to live and work. (Sears, Rowley, & Reid, 1990, p. 2)

Despite the fact that, in practice, public infrastructure investment is typically linked to its economic development impacts, the Sears et al. study concludes that all citizens should be afforded basic infrastructure services regardless of their economic development impact. However, as we suggested earlier, this study is the exception; the literature generally contributes to the dominant political practice of justifying investment in public infrastructure in terms of the expected economic development payoff—even though the payoff seems negligible by the results reported here.

A basic thesis of this chapter is that infrastructure investment to support health and safety of all citizens should not be withheld pending an estimate of its economic development impact. This is not to say that public infrastructure and economic development are not linked; rather, it is that public infrastructure stimulates and is at the foundation of the complete city—socially, spatially, as well as economic. A preoccupation with economic development at any cost threatens the health and safety of some citizens at worst and deprives others (rural and urban neighborhoods) of a level of indispensable infrastructure service that is the right of American citizens.

■ Spatialization of Infrastructure/Economic Development Decisions: Up With the "Edge" and Down With the Neighborhoods

The primacy of economic development as the policy context in which many governmental decisions are made has a profound impact on how legislative questions are cast. For example, support for providing day care is based on the increased *productivity of the work force*, rather than considerations for the parents or children. Thus, legislators are programmed to focus on the economic development aspect—productivity—and less on security and safety for the children and peace of mind for the parents. The impact of decisions regarding investment in infrastructure in this context reveals some unfortunate and, hopefully, unintended consequences.

For example, when infrastructure investment is made only in areas where there is some sort of measurable economic benefit in line with the studies discussed above, the spatialization of such investment can create some rather uneven geographies of development and reinforces ongoing spaces of decline in metropolitan areas. Joel Garreau (1991), in his book, *Edge Cities*, details the development of rapid growth at the edge of the metropolis—with substantial local and state investment in the infrastructure necessary to sustain such development. At the same time the urban cores of many of these areas continue to decline

(Byrum, 1991), the victims of a de facto policy of triage, where scarce resources for urban development are funneled to the places with the highest potential for a return—where the conditions of capital investment outlined by the economists in the previous section will be most clearly justified in terms of the projected *linkage* to economic development. Therefore, investment will be made in developable areas.

Garreau (1991) operationalizes edge cities the following way:

Has 5 million square feet or more of leasable office space—the workplace of the Information Age.

Has 600,000 square feet or more of leasable retail space.

Has more jobs than bedrooms.

Is perceived by the population as one place.

Was nothing like "city" as recently as 30 years ago.

Using this definition he was able to identify 205 edge cities or soon-to-become edge cities. Garreau finds that, in all but a few cases, central business districts contain a declining share of urban office space, as more and more of this developable space agglomerates at the edge. That is evidence enough that the phenomenon exists.

But what about the decline of the inner-city neighborhoods? Minneapolis and St. Paul are good examples. According to Byrum (1991), between 1950 and 1980:

The low income areas were depopulating while others were developing. The Minneapolis low income tracks depopulated from 56,900 people to 22,700 and the St. Paul low income tracks from 30,600 to 15,900, while suburban area developed from 9,700 to 46,000.

The suburban area was typified by home ownership. The low income tracks changed increasingly to rental housing and public housing.

In 1950 the median income of the high income tracks was 2.5 times that of the Minneapolis tracks. By 1980 it was 5.8 times larger. Income in the low income areas grew less rapidly than in the metropolitan area as a whole. In the high income area income grew more rapidly than in the metropolitan area.

This is obviously a decline in the core, and the citizens left behind cannot support rebuilding of their essential infrastructure. These data are probably attenuated in comparison to the trends in the older cities of the upper Midwest and Northeast.

Infrastructure investment will not be made in nondevelopable areas. It will not be made to maintain inner-city systems. These systems will deteriorate, forcing those who can afford it to leave the neighborhoods to move to the suburbs. Cleveland and Detroit are clear examples (Schorr, 1991). This, of course, erodes the tax base and reduces the property values in the city. Thus, the city has even less to invest in its infrastructure and finds itself having to choose between economic development, infrastructure, or other social programs. So far, economic development seems to win out as mayors fight to keep business interests in the city.

Given the unevenness in the findings regarding rural infrastructure and economic development, the conditions of spatial inequity in the impact of public works delivery is every bit as apparent in rural areas. There is no compelling reason to invest in rural areas. Where the driving force of successful service delivery is economic development, investment in critical rural services such as wastewater treatment is viewed as "nonessential" when compared to new age public infrastructure advances in telecommunications.

As a result, current policy can discriminate spatially by depriving entire, identifiable classes of citizens of vital services that protect and enhance their health, safety, and convenience. Those who benefit least from the public infrastructure-economic development linkage are citizens of core urban neighborhoods (African-American, Hispanic, poor) and distant rural areas (white, poor). Policies centered on stimulating economic development also disproportionately benefit definable classes— relatively affluent, predominately white people—the employable, mobile residents of suburbs and edge cities. The jobs created in the new central business districts downtown and in the industrial and office parks at the edge remain a great distance from the growing urban underclass—physically and socially. They just cannot get there.

The consequences of this situation are substantial. Urban infrastructure systems that are not maintained ultimately must be replaced at a cost far in excess of what it would have cost to maintain them in the first place. All of this occurs during a time when central cities are already experiencing an erosion of their fiscal capacity. Poor public transit systems do not enable urban residents to economically and conveniently travel to the edge cities for employment. All of these contribute to the erosion of urban life and peaceful coexistence of public citizens in the private city.

The longer run consequences of highways to suburbia enable residents to live away from the downtown, though they have access to it and, ultimately, to edge city. And to repeat a point made earlier, impact on the environment from the increased use of fossil fuels, coupled with traffic congestion, has an immediate impact on the residents of the core neighborhoods and a long-term impact on everyone else over time. The virtual disregard of incorporating health, safety, and convenience into the policy arena devalues the interests of the urban and rural poor, pitting their interests against those of developers and other allied interests.

■ Can Public Infrastructure Be Maintained in the Political Environment of Economic Development?

There are compelling reasons to maintain the integrity of all infrastructure systems. If the argument that the current policy blatantly discriminates against poor people's right to expect a decent level of health and safety and convenience from government is not forceful, perhaps the notion that there are hundreds of Los Angeles ready to ignite is. The unintended impact of the domination of private interests in the city should be heeded by policymakers. This does not mean that wholesale investment in all infrastructure to the exclusion of all else will solve potential urban and rural uprisings. As John Peterson said, "Not all infrastructure is equal" (Peterson, 1991). It is important to invest more, selectively, and smarter.

There is no agreement among experts on what level of annual infrastructure investment in existing physical systems is currently needed in the United States. The National Council on Public Works Improvement estimated $90 billion; the Associated General Contractors figured $141 billion; the Congressional Budget Office suggested $59 billion; and the Joint Economic Committee proposed $64 billion.[12] Surely there are more projects than there is available money to fund them all. There are physical reconstruction and maintenance needs that are pressing. What if we could make the available money more productive?

One way to do this is to invest in the research and development of more productive materials and processes for infrastructure construction and maintenance. For example, composite materials have been used in the space program and in the construction of commercial airplanes for years. Very few R&D funds in this country have gone to design and test the use of composites in construction. One promising research initiative is the use of pultruded fiber reinforced plastics in the construction of bridges. These lightweight bridges could be constructed off site (weather conditions would not influence the construction). They would then be conveyed via flatbed truck to the construction site and connected to the supports. Their sturdy though lightweight nature suggests obvious applications in emergency management (e.g., flooding) situations.

Current building codes call for bridges to contain reinforced steel bars for support—something made unnecessary by the new material. However, major barriers to the introduction of this new material, erected through archaic codes and the vested interests of the concrete and rebar industries as well as unions, must be overcome. New industries in composite materials could emerge, making construction more productive in the long run.

Another barrier to the use of more productive materials and processes is litigation. Architects and design engineers are unwilling to assume the liability for new materials and processes. That is why the government should fund demonstration projects in areas where building codes could be relaxed in the presence of scientific evidence that the new material or process is scientifically sound. The Committee for Economic Development views technology change as the *primary* source of economic growth in the United States (Melkers & Eberts, 1990). Research mentioned earlier suggests that the adoption of technological innovations in rural areas enhances their economic development. This being the case, such investment in R&D should make infrastructure construction and maintenance more productive, increase the number of firms engaged in producing the new materials, enhance economic development, and contribute favorably to the health, safety, and convenience of all. Consequently, more infrastructure dollars will be available to provide noneconomic-development-related infrastructure systems.

Availability of new technologies does not ensure their application in the field. The most frequently discussed example in this regard is potholes. For the most part, potholes are filled the same way they were filled in the 1950s, despite the fact that there are new materials and processes that are more durable. The combination of restrictive building codes and inadequate technology transfer has impeded change. In Ohio, public works practitioners report that they receive the overwhelming amount of information about new materials and processes from vendors and suppliers (who have a vested interest in their products), but they receive no help from universities and the research sector about the applicability of these products and the processes of change they should undertake to implement new products.

There are a number of reasons for the failure of the transfer of basic research to practitioners. First, there is no intellectual incentive for the basic academic researchers to either do applied research or engage in the development process.[13] The scholarly journals in engineering do not publish applied research. Public infrastructure officials rely almost entirely on the work of engineers to help them with new technologies and to design the mechanisms to transfer basic product change into new useable applications. The engineering community is simply not organized to accommodate these applications. Second, there is no existing clearinghouse for the information that does exist on technology transfer in the area of new product applications. Once again, what information centers do exist are typically operated out of engineering schools.

The development of applied uses for new materials and processes could be assisted by the National Science Foundation's Industry/University Cooperative Research Centers Program or its Small Business Innovation Research Program and the Army Corps of Engineers' Construction Productivity Advancement Research Program. The existing technology transfer aspects discovered for these products will need to be reviewed in light of the changing needs of public infrastructure, and the outmoded systems of codes and regulations presently in place. The result will be new mechanisms of technology application and institutional implementation.

In an era of fiscal austerity, where more and more responsibility for construction and maintenance of the infrastructure has fallen on state and local governments, the federal government can certainly support them by funding the kinds of research, development, and technology transfer efforts that will enable these governments to work smarter— introducing "smart" materials and better managerial practices. Some of the managerial practices include fast-tracking of projects, value engineering, and the creation of incentive/disincentive packages for contractors (Houlihan, 1992). This is a potential investment in the human infrastructure that designs, constructs, and maintains the physical infrastructure.

However, a weak economy exacerbates an unfortunate trend. Federal funds, no matter how useful they might be in this area, do not materialize. Overburdened state and local budgets get tight, and education, training, and travel budgets are cut (Felbinger, 1992). The ability of public works officials to take advantage of the new technologies is greatly reduced. City councils and state legislators must understand the value of formal education, continuing education, and conferences for their employees—especially managers—and especially in budgetary hard times. The case is particularly crucial for public works managers. Capital and operating budgets of their agencies are among the largest in most cities. With so many projects and systems operating, more efficient methods of operation and creative financing can save taxpayers millions, perhaps billions, of dollars on infrastructure projects. Unlike their counterparts of old, future public works managers do not have the luxury of learning management techniques hit or miss while spending years on the job. The changes are coming too fast and the technologies are too complex when dealing with expensive infrastructure systems. Constant maintenance of human infrastructure requires career-long attention to retooling or reeducation if public works are to be provided with balanced attention to a community's social as well as economic needs (Houlihan, 1992).

Can a community's full complement of public infrastructure programs be maintained in a political environment that places a primacy on projects servicing regional economic development? For those investments that directly contribute to economic development as the developers and politicians know it, definitely. Such projects contribute to the prevailing ideology of privatism in an era of mixed economic recovery. For those investments not so directly related to economic development, a case must be made that all citizens should expect a decent level of infrastructure investment to assure health, safety, and convenience. If the link cannot also be made between these concerns and those of the economy, the results for public works policy implementation will be decidedly more problematic. Investments in research, development, technology transfer, and human infrastructure development can ensure enough money for both in the long term through productivity increases and new materials' manufacturing.

■ Conclusion

A case has been made here that health, safety, and convenience of citizens should be the first consideration when making local investment decisions. Byrum (1991) argues that improving human development characteristics (health and safety) is more important than revitalizing

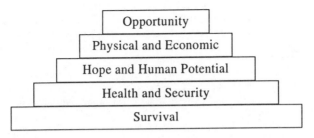

Figure 4.1. Building Blocks of Successful Neighborhoods
SOURCE: Byrum (1991, p. 84). Reprinted with permission from *Old Problems in New Times,* © 1992 by the American Planning Association, 1313 East 60th Street, Chicago, IL 60637.

areas in a marketplace sense—or at least it must precede it. He identifies a number of reasons why this is a key community responsibility:

> First, much of the work needed falls within the traditional city functions of public safety, public health, education, and maintaining overall neighborhood conditions.
>
> Second, failure to maintain these traditional services and other aspects of livability is a clear factor in the city decline pattern that has preceded us. Not letting this happen is to stop digging the hole.
>
> Third, the quality of life in inner city neighborhoods is critical to the future of the rest of the city. The better the conditions of our most difficult neighborhoods, the less negative impact they have on surrounding areas, upon the image of the entire city as a place to live and the strength of all city neighborhoods in the metropolitan housing market.

Byrum then posits a model of the building blocks of successful neighborhoods (Figure 4.1). I would argue that the model can be extrapolated to successful regions, both urban and rural. The fundamental building block is survival, followed by health and security. Without these, subsequent steps leading to overall economic development of a region do not proceed on a firm foundation, causing those development exercises to be at risk. The notion that economic development is a policy requisite should be rethought in light of these conclusions.

NOTES

1. See also A. Smith. (1976). "Of the expense of public works and public institutions" in *The wealth of nations* (Book IV, Chap. IX, Pt. 3, p. 723). Oxford, UK: Claredon.

2. The emphasis now in federally funded programs is to assist with the development of infrastructure revolving loan funds, rather than fully funding capital construction. In the past, however, there was some fiscal incentive for facility managers to allow their infrastructure systems to deteriorate beyond what could be fixed through routine maintenance (which is paid for locally) and wait until federal dollars could reconstruct them.

3. How this happened will be discussed below. The purpose to this point is to define the concepts of infrastructure and economic development and summarize the conditions that lead to their linkage.

4. Schultz (1989). Much of the historical material in this section relies on information from this source.

5. An engineer in the *Baltimore Sun*, April 24, 1905.

6. *Edmund Baker vs. City of Boston* (1831).

7. In *Watertown v. Mayo* (1972), this police power was restricted to cases where health and safety (explicitly not economic development) were both the explicit and implicit reasons for imposing the police power.

8. Of course some of this was tempered by Dillon's Rule, set out in *Merriam v. Moody's Executors* (1886), and later adopted by the Supreme Court in *Hunter v. The City of Pittsburgh* (1907). Dillon's Rule stated that municipal corporations may perform only those functions explicitly delegated to them by the states. The concept of "home rule" jurisdiction gets around the strict nature of Dillon's Rule.

9. Much of the information in this section was reviewed in Armstrong (1976).

10. Of course, some would argue that Davis-Bacon provisions undercut local government's ability to do projects efficiently because of higher wages. However, the spirit of the act is what is important here.

11. Highway expenditures have also been linked to gross state product: Garcia-Milla and McGuire (1987).

12. "Infrastructure: An Emerging Issue Needs Long-Term Solutions." (1989).

13. A proposed Ohio Infrastructure Institute would provide researchers with that incentive through research and demonstration project funding. The state legislature has not funded this venture.

REFERENCES

American Economic Development Council (AEDC). (1984). *Economic development today*. Chicago: Author.

Apogee Research, Inc. (1987). *Wastewater management: Current policies and future options*. Report prepared for the National Council on Public Works Improvement. Washington, DC: Advisory Commission on Intergovernmental Relations.

Armstrong, E. (Ed.). (1976). *History of public works in the United States: 1776-1976*. Chicago: American Public Works Association.

Aschauer, D. A. (1989). Is public expenditure productive? *Journal of Monetary Economics, 23*, 177-200.

Edmund Baker v. City of Boston, 12 Pick. 184 (Mass. 1831).

Bartik, T. J. (1985). Business location decisions in the United States: Estimates of the effects of unionization, taxes, and other characteristics of the states. *Journal of Business and Economic Statistics, 3*, 14-22.

Bingham, R. D. (1986). *State and local government in an urban society*. New York: Random House.

Bingham, R. D., Hill, E. W., & White, S. B. (1990). Preface. *Financing economic development: An institutional response*. Newbury Park, CA: Sage.

Birch, D. (1987). *Job creation in America*. New York: Free Press.

Blakley, E. J. (1989). *Planning local economic development: Theory and practice*. Newbury Park, CA: Sage.

Bluestone, B., & Harrison, B. (1982). *The deindustrialization of America*. New York: Harper & Row.

Boskin, M. J., Robinson, M. S., & Huber, A. M. (1987). *New estimates of state and local government tangible capital and net investment* (Working Paper Series, No. 2131). Cambridge, MA: National Bureau of Economic Research.

Byrum, O. (1991). *Old problems in new times: Urban strategies for the 1990s*. Minneapolis: Center for Urban and Regional Affairs.

Congressional Budget Office (CBO). (1988). *New directions for the nation's public works*. Washington, DC: Government Printing Office.

CONSAD Research Corporation. (1969). *A study of the effects of public investment*. Prepared for the Office of Economic Research, Economic Development Administration, Washington, DC.

Conservation Foundation. (1982). *State of the environment*. Washington, DC: Author.

Costa, J. d.S., Ellson, R. & Martin, R. C. (1987). Public capital, regional output, and development: Some empirical evidence. *Journal of Regional Science, 27*, 419-437.

Council on Environmental Quality. (1981). *Environmental quality*. Washington, DC: Author.

Deno, K. T. (1986). *The short run relationship between investment in public infrastructure and the formation of private capital*. Unpublished doctoral dissertation, University of Oregon.

Duffy-Deno, K. T., & Eberts, R. W. (1989). *Public infrastructure and economic development: A simultaneous equations approach* (Working Paper No. 8909). Cleveland: Federal Reserve Bank of Cleveland.

Eberhard, J. P., & Bernstein, A. B. (1985). A conceptual framework for thinking about urban infrastructure. *Built Environment, 10*, 254-268.

Eberts, R. W. (1990). *Public infrastructure and economic development* (Economic Review 10). Cleveland: Federal Reserve Bank of Cleveland.

Eisner, R. (1989). *The total incomes system of accounts*. Chicago: University of Chicago Press.

Eisner R. (1991, September/October). Infrastructure and regional economic performance. *New England Economic Review*, pp. 47-58.

Etzioni, A. (1983). *An immodest agenda: Rebuilding America before the 21st century*. New York: McGraw-Hill.

Felbinger, C. L. (1992). Public works in the U.S.A. In B. Houlihan (Ed.), *The challenge of public works management: A comparative study of North America, Japan, and Europe* (pp. 29-64). Brussels: International Institute for Administrative Sciences.

Forkenbrock, D. J., Pogue, T. F., Finnegan, D. J., & Foster, N.S.J. (1990). Transportation investment to promote economic development. In USDA, *Infrastructure investment and economic development* (Staff Report No. AGES 9069). Washington, DC: Agriculture and Rural Economy Division.

Fox, W. F., & Murry, M. N. (1989). *Local public policies and interregional business development* (Mimeo). Knoxville: University of Tennessee.

Garcia-Milla, T., & McGuire, T. J. (1987). *The contribution of publicly provided inputs to states' economies* (Research Paper 292). Stony Brook: State University of New York at Stony Brook.

Garreau, J. (1991). *Edge cities: Life on the new frontier.* New York: Doubleday.

Griffith, E. S. (1974). *A history of American city government: The progressive years and their aftermath.* New York: Praeger.

Hanson, N. M. (1965). Unbalanced growth and regional development. *Western Economic Journal, 4,* 3-14.

Helms, L. J. (1985). The effect of state and local taxes on economic growth: A time series, cross sectional approach. *Review of Economics and Statistics, 67,* 574-582.

Henning, D. H., & Mangun, W. R. (1989). *Managing the environmental crisis: Incorporating competing values in natural resource administration.* Durham, NC: Duke University Press. (See also, Council on Environmental Quality. 1970. *Environmental quality*)

Hite, J. (1989). *Financing infrastructure in rural America.* Paper presented at the Infrastructure and Rural Economic Development Symposium, Southern Agricultural Economics Association, Nashville, TN.

Holtz-Eakin, D. (1994). Public-sector capital and the productivity puzzle. *Review of Economics and Statistics, 76,* 12-21.

Houlihan, B. (1992). Conclusion. In B. Houlihan (Ed.), *The challenge of public works management: A comparative study of North America, Japan, and Europe* (pp. 341-359). Brussels: International Institute for Administrative Sciences.

Howe, F. C. (1967). *The city: The hope for democracy.* Seattle: University of Washington Press. (Original work published 1905)

Hunter v. The City of Pittsburgh, 207 U.S. 161 (1907).

Infrastructure: An emerging issue needs long-term solutions. (1989, July). *The Public's Capital, 1*(1), 1-3.

Jones, J. B., & Bachelor, L. W. (1986). *The sustaining hand: Community leadership and corporate power.* Lawrence: University Press of Kansas.

Logan, J. R., & Molotch, H. L. (1987). *Urban fortunes: The political economy of place.* Los Angeles: University of California Press.

Mann, P., & Beecher, J. (1989). *Cost impact of safe drinking water act compliance for commission-regulated water utilities* (NRRI Report 89-6). Columbus, OH: The National Regulatory Research Institute.

Markusen, A., Hall, P., & Glasmeier, A. (1986). *High tech America: The what, how, where, and why of sunrise industries.* Winchester, MA: Allen & Unwin.

Melkers, J. A., & Eberts, R. W. (1990). Can R&D be the R_x for the midwest? *Economic Commentary.* Cleveland: Federal Reserve Bank of Cleveland.

Mera, K. (1975). *Income distribution and regional economic development.* Tokyo: University of Tokyo Press.

Merriam v. Moody's Executors, 25 Iowa 163 (1886).

Munnell, A. (1991). *Comments on: "Is there too little capital?: infrastructure and growth" by Charles Hulten and Robert Schwab.* American Enterprise Institute Conference on Infrastructure Needs and Policy Options for the 1990s, Washington, DC.

National Council on Public Works Improvement (NCPWI). (1988). *Fragile foundations: A report on America's public works.* Washington, DC: Author.

National Research Council Board on Atmospheric Sciences and Climate. (1983). *Changing climate.* Washington, DC: National Academy Press.

Nordhaus, W. D., & Tobin, J. (1972). Is growth obsolete? *Economic Growth* [Fiftieth Anniversary Colloquium, Vol. 5]. New York: National Bureau of Economic Research.

Parker, E. B. (1990). Communications investment to promote economic development. In USDA, *Infrastructure investment and economic development* (Staff Report No. AGES 9069). Washington, DC: USDA.

Peterson, J. E. (1991). Infrastructure and growth: Some new thoughts. *Governing, 5,* p. 55.

Pisarski, A. E. (1987). *The nation's public works: Report on highways, streets, roads, and bridges.* Washington, DC: National Council on Public Works Improvement.

Reich, R. B. (1983). *The next American frontier.* New York: Times Books.

Rowley, T. D., Grigg, N. S., & Rossi, C. V. (1990). Water and wastewater investment to promote economic development. In USDA, *Infrastructure investment and economic development* (Staff Report No. AGES 9069). Washington, DC: USDA.

Schorr, A. L. (Ed.). (1991). *Cleveland development: A dissenting view.* Cleveland: David Press.

Schultz, S. K. (1989). *Constructing urban culture.* Philadelphia: Temple University Press.

Sears, D. W., Rowley, T. D., & Reid, J. N. (1990). Infrastructure investment and economic development: An overview. In USDA, *Infrastructure investment and economic development* (Staff Report No. AGES 9069, pp. 1-18). Washington, DC: USDA.

Smith, A. (1976). *An inquiry into the nature and causes of the wealth of nations.* Chicago: University of Chicago Press.

U.S. Congress, Office of Technology Assessment (OTA), based on Texas Transportation Institute. (1989). *Roadway congestion in major urban areas, 1982-1987* (Research Report). Washington, DC: Government Printing Office.

U.S. Congress, Office of Technology Assessment (OTA). (1990). *Rebuilding the foundations: A special report on state and local public works financing and management.* Washington, DC: Government Printing Office.

U.S. Congress, Office of Technology Assessment (OTA). (1991). *Delivering the goods: Public works technologies, management, and finance.* Washington, DC: Government Printing Office.

U.S. Department of Agriculture (USDA). (1990). *Infrastructure investment and economic development: Rural strategies for the 1990s* (Staff Report No. AGES 9069). Washington, DC: USDA, Agriculture and Rural Economy Division.

Village of Euclid v. Amber Realty Company, 272 U.S. 365 (1926).

Watertown v. Mayo, 109 Mass. 315 (1972).

Wellington, et al. Petitioners, 16 Mass. Pick. 87-106 (1834).

Wright, D. S. (1988). *Understanding intergovernmental relations* (3rd Ed.). Pacific Grove, CA: Brooks/Cole.

5   Public Infrastructure and Special Purpose Governments: Who Pays and How?

JAMES LEIGLAND

Collapsed bridges, closed roads, broken water mains, unreliable mass transit, and overcrowded prisons all offer eloquent testimony concerning the deterioration of America's public works infrastructure. A recent series of studies has confirmed that overall investment in the nation's public works has slowed significantly over the past two decades. As a result, the quality of the nation's public works is barely adequate to meet current needs and will be unable to meet the needs of future economic growth.

A 1985 study by the U.S. Department of the Treasury estimated that $1-2 trillion in new capital investment (in 1982 dollars) was needed to restart the process of adequately maintaining the nation's public works inventory. Such investment has increased substantially since the mid-1980s, but more recent studies confirm that the gap between required and actual capital spending is still widening (see Petersen, Holstein, & Weiss, 1993). Fiscal pressures on all levels of government have made it difficult for them to make what most experts see as the necessary, minimum levels of financial commitment to infrastructure. Federal budget deficits will continue to make adequate funding of public works repair and replacement a challenge for the president and Congress, as will statutory and constitutional limits on taxing, spending, and borrowing at the state and local levels. The effects of the Tax Reform Act of 1986 also continue to be felt by states and localities as constraints on public works financing.

A number of recommendations have been made to enhance the ability of public officials to finance and manage the repair and replacement of public works facilities. Wider use of earmarked taxes, user fees, and trust funds have been proposed, and the federal government has been urged to play a more active and consistent role in infrastructure capital investment. In early 1993 a Congressional commission recommended the formation of a national infrastructure corporation to help make infrastructure investments more attractive to investors.

A similar recommendation, which has received widespread attention since the early 1980s, calls on state and local governments to create their own government corporations to finance and manage public works services (deHaven-Smith, 1985; National Council on Public Works Improvement, 1988; Peterson & Miller, 1982; Porter, 1987; Porter, Lin, & Peiser, 1987). At the state and local levels, these government units are known as public authorities, special districts, boards, corporations, and so on and are said to offer ways of circumventing unreasonable restrictions on public borrowing, increasing the likelihood that construction management and service delivery are carried out more cost-effectively, and shifting the burden of public works financing away from all taxpayers to those directly served.

This chapter assesses the role of these kinds of special purpose governments in financing and managing the delivery of public works services. The discussion provides first an overview of what is known about special purpose governments and the nation's infrastructure crisis. Second, special purpose government participation in selected public works sectors is described. Third, broader policy issues raised by the role of special purpose governments in this area are discussed. A fourth and final section summarizes the paper and concludes with some recommendations regarding proposed wider use of special purpose governments.

■ **Overview: Special Purpose Governments and Public Works**

What Is a Special Purpose Government?

Special purpose governments exist at the state, interstate, and local levels, and constitute the fastest-growing, least well understood form of American government. Known across the country as authorities, districts, banks, services, systems, agencies, commissions, boards, associations, areas, companies, corporations, and the like, these entities participate in public works provision in a variety of ways: They build and run bridges, tunnels, toll roads, dams, ports, airports, railroads, and mass transit facilities. They provide essential services, including water supply, sewerage, irrigation, reclamation, and solid waste disposal. Over the past two decades, as bond banks, development authorities, and so on, they have increasingly come to administer a wide array of loans and subsidies to other governmental units engaged in public works activities.

A variety of definitions have been offered for these entities (or subgroups within the general category) by the U.S. Census Bureau, various social scientists, and state and local public officials. For the purposes of this study, the term *special purpose government* will refer to government units, most often known either as public authorities or special districts, created for limited public purposes and designed to operate outside the regular executive structure of state or local government. Most of these entities possess corporate identities separate from those of the "parent" governments, and issue debt backed by taxes or user fees.

Problems With Definitions and Data

No one knows with any precision what the role of special purpose government is in public works provision. This is largely because of the lack of any widespread agreement about definitions or types of special purpose governments, suggested above. There also has never been a comprehensive survey of these entities—no one knows how many are active across the United States. The Census Bureau counts only "independent special districts," and therefore does not collect information on special purpose governments that it classifies as "subordinate" agencies of regular government. As a result, census figures exclude hundreds and perhaps thousands of special purpose governments that are relied on heavily by regular government units to play leading roles in public works provision (Leigland, 1990).

A second reason for confusion about the role of special purpose governments in public works activities relates to confusion about what is and is not to be included as public works. There is no widely accepted single definition of public works or infrastructure; national studies are invariably based on different assumptions and include varying public works components. Some studies focus on all capital outlays by government and the private sector. Other studies narrow the scope of their inquiry considerably. One of the most comprehensive studies of public works needs in the late 1980s was undertaken by the National Council on Public Works Improvement (NCPWI, 1988). This research was limited to facilities for transportation (highways, streets, roads and bridges, airports, water transport, mass transit), wastewater treatment, water resources, solid waste disposal, and hazardous waste services. Some important categories of public works were not included, such as schools, public housing, power production, public buildings, and prisons.

Although most recent studies of infrastructure needs make use of different definitions and data, all of these studies reach similar conclusions about the fundamental need for public works services. In addition to studies by the NCPWI and the Department of the Treasury, cited above, other recent studies of infrastructure needs have been conducted by the Associated General Contractors of America, the Joint Economic Committee of Congress, the Congressional Budget Office, and the Congressional Commission on Infrastructure Investment. A recent review of these studies summarized the basic findings (Petersen et al., 1993):

- Total state and local government spending on core public works facilities is approximately $80 billion per year.
- To replace existing capital stock, meet performance standards, and handle new growth, spending must be increased by anywhere from $10 to $70 billion per year. Reasonable estimates that take into account inflation and other relevant factors suggest that annual public borrowing in support of infrastructure repair and replacement should reach a level of approximately $130 billion by the late 1990s, almost double the annual amount borrowed for these purposes during the late 1980s.
- The biggest infrastructure spending needs are in areas such as highways, roads, and bridges, with water supply and sewerage requirements following closely in second place.
- Regionally, needs vary, but investment for repair and new construction are almost equally important in older areas of the nation, as well as in newer, growing areas of the South and West.
- Some functional areas that have made heavy demands on public works spending in the past, such as public education, appear to require less attention in the future because of changing demographic profiles.

Indirect Evidence of a Large Role in Public Works Provision

Despite the lack of direct evidence regarding the role of special purpose governments in public works provision, it is possible to con-

TABLE 5.1 Usual Lead Roles in Public Works Categories

Categories of Public Works	Capital Financing	Ownership, Management, and O/M Financing
Federal-Aid Highways:		
interstate	F	S
non-interstate	S/L/PS	S/L
Non-Federal-Aid Roads	L	L
Airports:		
major commercial	L/PS	L/PS
general aviation and smaller commercial	F	L/PS
Airways	F	F
Mass Transit	F	L/PS
Water Supply (Urban)	L/PS	L/PS
Wastewater Treatment	F/S	L
Water Resources:		
ports (landside)	L/PS	L/PS
dredging/navigation	F	F
major dams	F	F
urban stormwater	L	L
nonfederal dam safety	PS/L	PS/L
Solid Waste	L/PS	L/PS
Hazardous Waste:		
currently generated	PS	PS
site cleanup	F	PS

NOTE: F = Federal, S = State, L = Local, PS = Private Sector
SOURCE: National Council on Public Works Improvement (1988).

clude from available indirect evidence that these entities are responsible for a large percentage of the financing and management activities attributed to state and local governments. Because of relatively small roles traditionally played by the federal government and the private sector, state and local governments have been credited with as much as 90% of all current public works spending (Petersen et al., 1993).

Table 5.1 indicates the usual lead roles in the financing and management of various kinds of public works facilities, as defined by the NCPWI. The principal role of the federal government is in the capital financing of interstate highways, general aviation airports, mass transit, major dams, and so on. Its traditional role in capital financing of wastewater treatment is being phased out. The federal government plays almost no role in the ownership, management, or operations and maintenance financing of the kinds of public works services listed. The NCPWI study confirmed this conclusion with information on the changing size of total federal public works spending relative to state and local spending since 1960. In constant dollars, the overall size of total government spending has grown over this period, but federal spending has declined since the late 1970s, as has the federal share of government spending in this area.

As Table 5.1 indicates, states usually play lead management or financing roles in only a few of the public works categories listed. The principal exception to this lack of involvement is in the area of federal-aid highways (primarily non-interstate), on which states spend approximately 80% of their infrastructure funds, typically raised via state gasoline taxes. But nationally, such roadways are dwarfed in size by locally provided roads, which make up more than 75% of total roadway mileage. State roles in financing mass transit and wastewater treatment are growing, as the federal government continues to reduce its participation in these areas. As indicated in Table 5.1, local governments have primary financing and management responsibility for local roads, airports, water supply, port facilities, urban storm water facilities, and solid waste disposal. Local governments also own and operate mass transit facilities. In many localities, service roles are shared with the private sector. Unlike federal and state governments, local governments have steadily increased both the constant dollar amount of total public works spending from the early 1970s to the late 1980s, and their overall share of total government spending. By the late 1980s, local governments accounted for nearly 50% of all public works spending.

Much of the spending attributed to state or local governments has actually been carried out by special purpose government units, acting as subsidiaries of regular state and local government. Although Census Bureau statistics are incomplete, they point to a large role for special purpose governments in this spending. For example, using 1982 as a base year, the bureau's statistics show that the amount spent by all special purpose governments on public works (as defined by the NCPWI) was $13.06 billion (a 71% increase over the 1977 figure). A 1990 study added to this amount similar expenditures by special purpose governments excluded from Census Bureau totals as subordinate agencies. The result shows that total spending by all special purpose governments in 1982 actually exceeded $21.2 billion, or approximately 33% of total public works spending by state and local government (see Leigland, 1990).

By 1995, taking into account new Census Bureau statistics, the continued exclusion of entities defined by the bureau as subordinate agencies, additional types of public works provision, and the growth in levels of regular government spending, it is likely that special purpose government spending on public works may exceed 43% of total state and local government spending for such purposes. Assuming that states and localities account for 90% of all spending on public works, this puts the national amount attributable to special purpose governments at approximately 39%.

Financing Public Works

The reason for this large and apparently growing role of special purpose governments in public works provision has to do primarily with their ability to effectively finance construction and renovation of these kinds of facilities. All levels of government face the traditional choice of paying for large public works projects via "pay-as-you-go" appropriations, debt financing (or "pay-as-you-use"), and various forms of lease financing (the choice of course depends in part on options allowed by state law). Growth in the level and complexity of state and local debt issuance has occurred largely because borrowing has emerged as an important method for financing large public works projects. It is estimated that bonds finance 40% of new annual infrastructure investment (NCPWI, 1986, p. 69).

It is likely that considerably more than half of this revenue debt is issued by special purpose governments, although no precise data has been accumulated on this subject. The advantages of using these kinds of government units to issue debt for public works financing are well known to state and local officials. Debt or tax ceilings and referenda requirements may limit the ability of regular governmental units to finance desired projects. Financial aid (e.g., from the federal government) often encourages the formation of new special purpose governments, and this aid may then be used to leverage additional financing. Federal assistance was instrumental in the creation of drainage, sewage treatment, transit, flood control, irrigation, and many other kinds of districts and authorities. As semi-independent corporate entities, exempt from many regular government procedural rules, special purpose governments usually can avoid many of the restrictions inherent in the normal government budget and appropriations processes.

Revenue debt restricts internal financing flexibility somewhat because project revenues are usually segregated and unavailable for general-fund purposes. Interest costs for revenue debt also tend to be higher because of the somewhat riskier nature of the sources of repayment: Revenue bond issues sometimes require "credit enhancements," such as bank letters of credit, surety bonds from insurance companies, municipal bond insurance, or pledges of state financial support for debt service payments. But proponents of revenue bond financing argue that by and large, these limitations are more than balanced by the flexibility afforded by this type of financing: It allows debt issuers to bypass statutory or constitutional limits on tax-supported debt, avoid the need to have voters approve "full faith and credit" bond issues (as most general purpose governments using tax-based financing are required to do), and is consistent with demands that the cost of public projects be borne by project beneficiaries.

User charges and lease payments are the two traditional types of project revenues used to secure revenue bonds issued by special purpose governments, but other kinds of revenue may also be pledged in support of such debt. For example, special purpose governments in some states may pledge special assessments as security. In California, courts have ruled that special assessments are not taxes and therefore are not subject to the voter approvals and debt limit restrictions that normally affect tax-based financing in that state.

Government subsidy, particularly federal aid, plays a significant role in the capital financing activities of many kinds of special purpose governments. For example, the Federal Aviation Administration provides funding to many airport authorities and districts for the development of capital plans, and the Army Corps of Engineers shares responsibility for port development with port authorities and districts. Until the early 1980s the federal government was committed to providing up to 55% of the total capital costs of qualified communities' sewer systems, under the Clean Water Act and its amendments.

Legal and Administrative Considerations

Even when borrowing is a minor consideration, special purpose governments offer substantial advantages in providing public works services. User fees provide revenue for approximately 75% of public works capital spending, and approximately 50% of revenues spent on operations and maintenance (NCPWI, 1988, p. 11). Because of their freedom from routine government oversight of spending decisions, special purpose governments typically have more flexibility than units of general purpose government in deciding how user fees will be spent. These entities are also typically more capable of creatively designing and using effective alternatives to user fees. One variation enjoying growing popularity among special purpose governments is the capital recovery (or "development" or "impact") fee. These are set or negotiated charges imposed on developers for infrastructure expansion or construction. Exactions are facilities built by developers and dedicated to public purpose. Although the size of capital recovery fees may be large, recent studies indicate that they cover substantially less than half of the direct costs of providing public works.

Finally, the use of the special purpose government mechanism is also associated with a variety of legal, administrative, and political advantages. In actual practice these perceived capabilities may genuinely enhance the financing strengths of special purpose governments, but they are often discussed as if they are reasons in and of themselves for the use of this mechanism. Legal and administrative factors include the ability to deal with service areas that do not coincide with those of existing general purpose governments (i.e., smaller than a rural county or township, or larger that single government jurisdiction). Special purpose governments are also thought to be better able to provide services that are administratively complex or entrepreneurial, requiring marketing and procurement flexibility. Administrative flexibility may also include freedom from regular government regulations on procurement or civil service. Sometimes, special purpose governments are created with the express purpose of keeping a function free from partisan political influences. In general, however, the largest impact of this legal and administrative freedom and flexibility seems to be in attracting underwriters and investors to the debt issues sold by these entities.

■ Special Purpose Government in Public Works Sectors

Special purpose governments address the problems of financing the provision of public works in a variety of ways, depending largely upon the particular public works sector in which the entities operate, and the specific reasons for which the entities have been created. The following sections discuss the roles played by these entities in selected public works sectors, including transportation (highways, roads, bridges, and tunnels, mass transit, airports, seaports), water supply, wastewater treatment, and solid waste disposal (for additional, recent information on these sectors, see Peters, Lovette, & Choe, 1993).

Highways, Roads, Bridges, and Tunnels

Highways, roads, bridges, and tunnels constitute one of the most critical of all public works sectors. The nation's highway and roadway system is 3.9-million miles long and serves 187 million vehicles per year. In 1989, $65 billion was spent on the system, including capital improvements, maintenance, and operating costs incurred at all levels of government. But current spending is not enough to meet the system's needs for repair and replacement. In 1989 the federal government estimated the cost of repairing the existing backlog of highway deficiencies at $400 billion. The cost of congestion in the nation's 39 largest metropolitan areas was estimated at $34 billion in 1988 (see U.S. Department of Transportation, Federal Highway Administration, 1991). Numerous special purpose governments exist across the country for the construction, management, and maintenance of highways, roads, bridges, and tunnels. These entities present tremendous variety in terms of organizational forms and financing techniques. The major toll road authorities account for most of the major toll-supported highway construction, operation, and maintenance in the United States. Some highway districts, such as those in California and Georgia, also collect tolls. Toll bridge authorities are used in South Carolina, New York, Vermont, Missouri, New Jersey, Alabama, Pennsylvania, New Hampshire, Arkansas, Colorado, and Virginia.

Road districts, which raise revenues via benefit assessments as well as tax levies in some cases, exist in a large number of states, often as dependent entities that provide services or collect money under the administrative and financial control of parent government entities. Many independent road districts exist as well, and may be called road improvement districts (Arkansas, Nevada), county paving districts (South Dakota), road assessment districts (Oregon), good road districts (Idaho), and so on. Road improvement districts in Idaho, South Dakota, Oregon, and Arkansas may levy property taxes. In general, the financing needs—and challenges—in this area are among the greatest faced by all types of local governments. Over the past 30 years, the primary state source of transportation revenues has been motor fuel taxes. These taxes now account for about half of all state transportation revenues. Today, the principal revenue sources of special purpose governments that build and maintain highways, roads, bridges, and tunnels are usually identified as user charges (other than fuel taxes), such as tolls and charges on benefiting properties (such as special benefit assessments or tax increments). Special benefit assessment is a financing technique that is usually applied to road or street improvements, often in unincorporated areas not otherwise serviced by a metropolitan department of transportation (although assessment districts are now being used more frequently in cities). Whereas turnpikes are typically operated by independent districts or authorities, benefit assessment districts are often dependent entities, in the sense that they exist under the administrative and financial control of a parent entity.

Mass Transit Administration

There are some 1,800 public mass transit systems in the United States. Most are not self-sustaining enterprise activities, and in this sense mass transit is different from the kind of function traditionally identified with special purpose governments. Nevertheless, most mass transit passenger miles are traveled on facilities owned and operated by public authorities or special districts (more than 50% of those miles are traveled in the Chicago and New York City metropolitan areas alone). Table 5.2 lists the 10 largest state systems for financing and managing urban mass transit—each of the indicated systems is centered around one or more special purpose governments.

The legislatures of a number of states have authorized, through enabling legislation, the efforts of local governments to form authorities or districts to provide mass transit services. For example, voters and legislative bodies in any county in Wisconsin may form a metropolitan transit authority. In Michigan and Indiana, metropolitan transportation authorities or regional transportation authorities may be created by legislative initiative of participating counties. Cities in Indiana may also create public transportation corporations. Cities and/or counties in Ohio may create regional transit authorities. Iowa cities, individually or in partnerships, may form transit authorities. Local mass transit districts may be created in Illinois by referendum in one or more of the cities or counties serviced by the district. The most important source of

TABLE 5.2 The 10 Largest State Urban Public Transit Systems

State	Principal Public Transit Operator
New York	Metropolitan Transportation Authority (Various subsidiary authorities) Port Authority of NY and NJ (PATH)
California	San Francisco Bay Area Rapid Transit District Southern California Rapid Transit District (Various other transit districts)
Illinois	Chicago Transit Authority Regional Transportation Authority (Various mass transit districts and urban transportation districts)
Pennsylvania	Southeastern Pennsylvania Transportation Authority (Various metropolitan transportation authorities)
Maryland	Washington Metropolitan Area Transit Authority
New Jersey	New Jersey Transit Corporation
Massachusetts	Massachusetts Bay Transportation Authority
Ohio	(Various regional transit authorities)
Texas	(Various metropolitan rapid transit authorities)
Michigan	Southeastern Michigan Transportation Authority (Various metropolitan transportation authorities and mass transportation authorities)

SOURCE: Derived from data presented in Leigland (1993).

mass transit revenues is direct operating or capital assistance by federal, state, or local governments. Additional funding may come from a dedicated sales tax; Atlanta, New Orleans, and Austin use this technique. The New York City Metropolitan Transportation Authority (MTA) has received budget surpluses from the Triborough Bridge and Tunnel Authority and some of the latter's net revenues have been pledged as security for MTA revenue bonds. By the late 1980s less than one third of the total amount spent on mass transit came from transportation operating revenues.

Most transit operating revenues come from fares, but a small and growing percentage of revenues is derived from the use of a variety of innovative financing techniques that have been experimented with over the past decade by special purpose governments, including connector fees, negotiated investment, benefit assessments, joint venture approaches, and so on.

Airport Administration

By the late 1980s more than 450 million passengers were being served by the nation's system of airports. That figure is expected to increase to one billion by the year 2000. The number of seriously congested airports is expected to more than double, to 58, by the end of the century. Special purpose governments are increasingly being used by state and local officials to meet the financing and management needs of airports under their jurisdictions. A survey conducted by United Airlines in the late 1980s found that 12 of the nation's 25 busiest airports were managed by special purpose governments. Across the country, these entities are known as boards, commissions, agencies, authorities, corporations, trusts, and of course, simply airports.

Twenty-three states have passed general enabling laws to allow for the creation of airport authorities or districts, usually by the governing boards of one or more local government entities. In Illinois, a circuit court judge can create airport authorities on petition of voters and after a hearing and local referendum. Florida, Maine, Nevada, Virginia, and South Carolina have all created airport authorities by special legislation. Airport authorities or districts, like municipal airports, typically derive revenues from terminal rentals and landing fees, usually secured by special contracts, known as *use agreements,* with airline carriers. Concessions and other revenues generated by the airport facilities are also secured by these contracts. Airport operators also may receive grant-in-aid planning and development funds from Federal Airport and Airways Trust Funds.

As the largest source of airport revenues, airline use agreements also typically secure most airport revenue bond issues. In these kinds of financings, the reaction by investors (and the cost of borrowing to the issuer) is based largely on an analysis of the creditworthiness of the airlines involved.

Sea and River Port Administration

More than 130 domestic seaports currently provide a variety of port-related facilities in the United States. Most of these facilities, including almost all large port operations, are managed by special purpose governments. However, some entities, known as port districts or authorities, have no sea or river port responsibilities at all. These entities have traditionally carried out economic development activities in many communities, acquiring land, financing and/or constructing buildings, installing capital equipment, selling or leasing these properties to private businesses, and issuing industrial development revenue bond debt. The largest and most financially secure special purpose governments involved in port management account for most of the port-related revenue bond financing carried out over the past two decades. Those authorities operate port facilities in New York City (the Port Authority of New York and New Jersey), Boston, Jacksonville, Los Angeles, New Orleans, Long Beach, Oakland, San Francisco, Portland (Oregon), Seattle, and Tacoma, Washington.

Principal revenues from port operations include docking and loading fees, terminal facilities rentals, concessions, and so on. Facilities for handling containerized cargo have become extremely important for attracting and keeping shipping business at most large ports, and many U.S. ports now handle containerized as well as bulk cargo. Smaller ports that handle bulk cargo exclusively have found themselves at the economic mercy of volatile markets for such bulk commodities as coal and fertilizer. Ports also derive revenues directly from specialized services and facilities, and indirectly from the increased business in shipping that such specialized facilities often provide. Port authorities and districts with a variety of different kinds of operations typically find it much easier to carry out capital financing than authorities with seaport responsibilities only. Many of these large, multifunction entities now make use of pooled bond reserves, a technique pioneered by the Port Authority of New York. Pooled reserves involve the establishment of a single common bond reserve fund into which each project's revenues are deposited. All bonds are repaid from that fund.

Water and Sewer Services

By the early 1990s many different types of special purpose governments were offering sewerage and water supply, and related services—it is almost impossible to generalize about this form of special purpose government. Many states authorize the formation of several different types of water districts or authorities under general law. In addition, many entities have their own distinctive and characteristically complex statutory charters. Water authorities or districts may exist as rural irrigation or improvement districts, common in the West and Far West, which serve as retail marketers of water obtained from both local and distant sources. They may also act as water wholesale collection systems, involved in the joint development of a common supply of water resources. In these cases, the wholesaler (a district or authority) becomes a borrower on behalf of smaller entities, which contract to assume shares of the wholesaler's operating costs and indebtedness. The largest wholesalers are active in the debt markets, and typically enjoy reputations as strong credits.

Sewer authorities or districts range from small rural entities to huge regional wholesale entities that construct and operate regional treatment facilities. The latter often provide wholesale treatment services while municipalities operate and maintain local collection systems for commercial, industrial, and residential use. Most states have chosen both to expand the purposes of existing special purpose governments that provide water and sewer services, and to create new types of special purpose entities to serve increasingly large geographic areas and fill a variety of purposes. The new type of entity may, in fact, encompass all or part of existing districts or authorities (as well as other water supply areas). This results in layers of special purpose entities and a patchwork of authority over water services in the same area. Development districts constitute a type of entity in this general category that has proliferated over the past 30 years. They are usually created in much the same way as western special water districts—through petition by a majority of the landowners or, depending on the jurisdiction, by the landowner or owners who hold enough land to constitute more than half of the assessed value of the land in the area. This type of entity is strongly associated with land development and can most frequently be found on the fringe of urban areas.

Water and sewer districts, authorities, commissions, and the like are among the financially strongest of all special purpose governments. Like public power utilities they are usually monopolies, typically with almost complete control over rate making. Basic revenue streams are usually derived from user charges and connection charges. Sometimes these entities are able to have local ordinances passed to require mandatory connections to the system by all residents. In some jurisdictions, water quality standards by the local board of health may in effect require connections. In many cases, sewer charges are based on water usage. Water and sewer revenue bonds—especially those issued by special purpose governments with significant operating histories (5 years or more)—are considered among the safest municipal investments and are relatively easy to market.

Solid Waste Management

Special purpose governments established for the purpose of providing for solid waste management services may be authorized to provide for the collection and disposal of solid wastes and/or authorized to construct, operate, and maintain solid waste treatment and disposal facilities, including resource recovery facilities and facilities that burn solid waste to produce energy. They may be established under state law, by ordinance of one or more contiguous municipalities (e.g., in Iowa, Kentucky, Louisiana, Minnesota, Montana, New Hampshire, New Jersey, New York, Vermont) or on petition of electors (e.g., in Illinois). Approval of the voters may be required (e.g., in Illinois, Louisiana, Rhode Island). The governing board may be appointed (usually by the establishing body) or elected. These entities may be known as garbage districts, solid waste management districts, solid waste disposal districts, or regional refuse disposal districts or authorities. There are a number of distinctly different sorts of special purpose governments involved in solid waste disposal. The first is the garbage district (though it may operate under other names), which is likely to be a local government taxing area rather than an independent entity. A second form more likely to issue municipal debt is a district or authority with a resource recovery project as part of its program.

Garbage districts generate revenues from service charges or benefit assessments; sometimes these revenues are used to pay private firms who contract with the district to provide pick-up services. Waste-to-energy projects may derive a portion of their revenues from the production and sale of energy. Revenue bond financing for new recovery projects must be supported by independent feasibility studies indicating the likelihood of revenues necessary to make the project economically feasible. A number of communities have faced legal problems as a result of their support for waste-to-energy projects undertaken by independent districts or authorities.

■ Policy Implications of the Role of Special Purpose Government in Public Works

Changes in the Municipal Bond Market

Borrowing by state and local governments of all kinds has been a major and growing source of public funds for more than three decades.

The Census Bureau reports that long-term capital raised on the tax-exempt bond market increased from $8 billion in 1960, to $47 billion in 1977, $218 billion in 1985, and $230 billion in 1992. The growing financial role of special purpose governments is due to state laws that are restrictive with regard to borrowing and spending by general purpose governments, and liberal with regard to the creation, borrowing, and spending of special purpose government units. The growing role of these entities has also been influenced by the financial advisers, bond attorneys, underwriters, and brokerage firms who have in many cases preferred to work out bond security arrangements with organizations separated administratively from general government.

The growth in municipal borrowing has been accompanied by several pronounced trends. One is the shift from general obligation (GO) bonds to revenue bonds. Most full-fledged GO bonds backed by "full faith and credit" (usually taxing power) are issued by general purpose governments, although some special purpose governments issue them. Most revenue bonds are issued by authorities and districts. In 1970 revenue bonds accounted for about 30% of long-term tax-exempt borrowing. By 1990 revenue bonds accounted for more than 75% of long-term tax-exempt borrowing. Clearly more than half the total national dollar volume on the tax-exempt market has been issued by special purpose governments. This raises serious questions about the possibility that borrowing by special purpose governments may to some extent crowd out general state and local borrowing, by raising interest costs that regular government entities must pay, and by taking a larger share of the debt that the market will absorb from the territory of each state and locality. The effect is evident in years of high interest rates and stable volume. Thus, for example, in 1981 more than $1 billion in planned state and local issues were postponed or canceled. The effect is hard to identify during years of falling interest rates and rapid growth of total volume. But the crowding out effect does appear to exist to some extent.

The second major trend that accompanied rapid debt volume growth from the early 1970s to the mid-1980s was the shift in purposes of expenditure away from traditional government infrastructure and services to private and mixed purposes. This shift accentuated the trend from general government to special purpose governments, because the latter were not spending most of the funds they raised on the maintenance, replacement, and construction of roads, bridges, rolling stock, water and drainage facilities, schools, and other well-documented infrastructure needs. More than half the total money raised through tax-exempt borrowing prior to the Tax Reform Act of 1986 was channeled into private housing mortgages, loans to businesses and developers, student loans, private health and university systems, industrial pollution control equipment, and leased buildings, stadiums, and community centers. Hence, the rising volume of tax-exempt borrowing probably did not fully compensate for declining federal aid for local public works infrastructure.

The effectiveness of the Tax Reform Act of 1986 (TRA) in reversing this trend and channeling borrowed funds back into public infrastructure is still difficult to assess, because such benefits do not clearly outweigh other costs associated with TRA. Such costs have resulted from the need to reduce in size or eliminate some infrastructure projects—or finance them with taxable bonds—in order to meet new state limits on tax-exempt borrowing. Other costs have involved new limits on arbitrage and advance refunding, and new, extensive limits on public-private partnerships used to provide infrastructure services. Some of the Clinton administration's first proposals in 1993 to facilitate state and local infrastructure financing called for exemptions of certain kinds of public borrowing from the restrictive state bond caps created by TRA.

Impact of Bond Financing on Project Viability

Special purpose governments use several kinds of revenue to secure bonds and to pay back principle and interest over time: taxes, special assessments, fees and user charges, investment income, grant funds from state and local government, lease payments from state and local government, contract payments from state and local government, debt service supplements—when necessary—from appropriations, and earmarked taxes levied by state or local government.

One policy question is whether the discipline of issuing bonds through the private investment market assures economic feasibility and financial responsibility. The answer is sometimes, but not always. First, the financings by small special purpose governments are often privately placed with local banks and are subject neither to the rating evaluations carried out by the national rating agencies, nor to the elaborate bond resolution and prospectus disclosure process. The larger authority and district financings are subject to those processes.

Bond defaults and near defaults of the 1970s and 1980s have highlighted the fact that revenue-secured borrowing by special purpose governments is indeed riskier than borrowing by general purpose governments. Those events enlightened the public to some extent about the abuses possible with debt financing of capital construction, and forced the investment community to scrutinize more closely revenue bond issues sold by these kinds of government entities.

For example, the Advisory Commission on Intergovernmental Relations (ACIR) reported that from 1972 to 1983, a total of 36 local governments defaulted on outstanding debt. Of that total, 25 defaults were by special purpose governments that had issued revenue bonds (ACIR, 1986). The short-term default of New York State's Urban Development Corporation (UDC) triggered statewide financial problems during the mid-1970s, as well as criticism by the Securities and Exchange Commission of the two major credit rating firms and the Wall Street community in general for a lack of due diligence in recognizing evidence of UDC's deteriorating financial condition (see Walsh, 1978). The $2.25-billion default of the Washington Public Power Supply System (WPPSS), in 1983, was the largest single public financial collapse in history (see Leigland & Lamb, 1986). Both of those financings, and several highway defaults, have involved highly rated bond issues subject to extensive credit analysis, and strongly recommended by acknowledged investment community experts.

Impact of Bond Financing on
Infrastructure Maintenance

In general, the tax-exempt bond market has made substantial contributions to new construction but has a built-in market bias against new fundings for repair and rehabilitation. Analyses of new bond issues in areas relating to water, waste disposal, and transportation confirm findings from past literature that revenue bonds are usually issued by special purpose governments for new construction (or debt refunding). Major efforts directed at repair and rehabilitation are much more often funded through general government GO bonds, such as state transportation bonds for highway repair. A related question is whether special purpose governments are more likely than general government agencies to keep existing facilities in good repair out of operating revenues. No general answer to this question is possible. Trust indentures, for bond issues that are placed through national and regional underwriters for borrowing secured by operating revenues, do usually specify flow of funds for debt reserves. Often these specifications use gross revenues. More recently the importance of using net revenues for those provisions has been recognized, considering the importance of operating and maintenance costs to be paid for the organization to remain viable. In any case, the maintenance and rehabilitation record of special purpose governments ranges from excellent to poor, as it does for cities. Some cities are noted for effective sinking fund and maintenance operations; others, particularly those with a history of budget problems, have clearly deferred maintenance, and their public works facilities have suffered from disinvestment. Similarly, the facilities of some special purpose governments have suffered as a result of deferred maintenance. It is particularly difficult politically for small special purpose governments to raise rates or taxes for repair and rehabilitation that will not result in increased user fees or improvements that are readily visible to taxpayers. Many urban fringe districts are annexed before that point is reached, and the maintenance as well as remaining debt service fall to city or county governments.

Many metropolitan or regional authorities or districts for which revenues have been declining and maintenance deferred (e.g., transit and some water and highway entities) do not have access to the bond market without backing of new earmarked tax revenues or other state or municipal tax sources. In contrast, recent billion-dollar capital plans of the Port Authority of New York and New Jersey have been devoted largely to improvement of existing tunnels, bridges, airports, and rail transportation facilities. Special purpose governments that have not deferred maintenance are those with strong and independent sources of finance. Their earmarked revenue streams are protected from cutback in times of budget scarcity. It is also the case that, for multipurpose public authorities and districts, the record of maintenance has been better for the income-generating projects or services than for the deficit-producing services (e.g., rail transit). Earmarked taxes associated with bond issues (e.g., gas taxes dedicated to a transit authority) are safer protection for operating budgets than annual appropriations. Special purpose governments dependent upon operating subsidies from general government, including sewage treatment and transit, have been more vulnerable to cutbacks. It may be considered a policy alternative to insulate the operating revenues of certain public works from public budget options. But that alternative does involve a sacrifice of the ability to apply broader priorities and tradeoffs through a budget process.

In any case it is clear that both project financing plans and audits should ensure that budget operations and maintenance by special purpose governments cover the life of a capital facility. In addition, work is still needed on maintenance, replacement, and design standards for public works projects of different types.

Self-Financing and Intergovernmental Aid

The argument in favor of wider use of special purpose governments (and expanded roles for existing entities) suggests that their separate administrative existence makes possible financial independence and helps realize their tendency toward efficiency based on reliance on user charges, particularly marginal price calculations. The empirical questions are: To what extent are special purpose governments self-financing, and what mix of revenues and pricing methods are they using?

Census Bureau data indicate that special purpose governments classified by the bureau as "independent special districts" active in public works fields are predominantly not self-financing. Census figures indicate that for such entities, taxes and charges accounted for less than 50% of total special district revenues during the late 1980s. The largest portion of the other 50% was accounted for by intergovernmental aid and has as its source general government taxes. According to the Census Bureau, these funds grew from 30% of special district revenues to 50% in the decade. Cutbacks in federal grants-in-aid have slowed this growth over the past 10 years. Within these aggregate figures range organizations from the Port Authority of New York and New Jersey, which issued a 10-year, $5-billion infrastructure financing plan in the late 1980s, and is essentially self-financing, to small sewage treatment districts or authorities mainly funded by federal grants and state and municipal governments. One major issue has been how to move profits from airport and automobile transportation services, and real estate assets of the large authorities and districts, into the typically deficit-producing and maintenance-deferred public works, such as rail transportation and non-toll highways. These transfers, when they have taken place, occur after protracted legal and political controversy. Examples include internal subsidization of the trans-Hudson rail line and the infrastructure bank by the Port Authority of New York and New Jersey; the New Jersey Transportation Trust Fund Authority, partly capitalized by funds from the state's toll road authorities; and internal subsidization of rail services in the MTA of New York by consolidation that brought the Triborough Bridge and Tunnel Authority under its umbrella.

The many types of dependence by special purpose governments on subsidy from regular government entities defy generalization. Federal aid has played a larger role in revenues of districts, authorities, and the like than in the revenues of any other kind of local government. Cities have been the next highest in dependence on federal aid. (Federal aid, like bond finance, has been largely concentrated on capital expenditures, thus contributing to a system bias against maintenance and rehabilitation, and sometimes causing serious shortfalls in operating revenues, for example, in sewage treatment plants.)

State aid has traditionally been a smaller, but more stable source of assistance; but more state assistance comes in the form of subsidies for services specifically identified as meeting social objectives and externalities. In addition, unplanned and largely uncounted state subsidies are caused by authority or district underestimates of costs or overestimation of demand, by political limitations on rate increases, and by various tax exemptions and land use activities by special purpose governments.

User Charges and Pricing Systems

To the extent that special purpose governments do rely on user charges, what kind of pricing systems do they use? A brief review of the data and the case study literature indicates that pricing calculated on marginal costs and peak load price increments, two elements that should theoretically induce efficiency, are extremely rare. It may be the case that where such pricing mechanisms have been applied (e.g., in metered water services), they are as common in municipal water systems as in district and authority water systems. This is a topic that requires further study. The widely accepted belief that special purpose governments rely heavily on self-financing assumes the existence of pricing mechanisms about which little empirical information has been gathered. Most of the available case studies indicate that user charges are commonly calculated by two methods: (a) average estimated costs to the entity—simply taking total estimated budget less expected subsidies and allocating them to units of output, users, or assessees; or (b) when costs regularly exceed operating revenues, pushing for maximum politically feasible rate increases, and developing other sources of internal revenues (moving into more profitable service areas) or demanding political subsidies to make up the difference.

The implications of these pricing systems for theoretical efficiency have not been completely analyzed. Many kinds of dedicated district or authority revenues (including earmarked taxes and average-price below cost user charges) do appear to increase revenue streams for public works, but they do not encourage efficiency in use. Theories of efficiency posit that if marginal costs are below average costs—as in the case of high capital costs and low operating costs for treatment plants— "efficiency calls for an operating subsidy paid out of general revenues" (Vaughan, 1983, p. 50). The empirical evidence on special purpose government financing has not been analyzed in terms of this principle.

Additional Revenue Sources and Financing Techniques

Special purpose governments of various kinds play important roles in exploiting new sources of revenue and new kinds of financing techniques used in connection with public works provision.

Special Benefit Assessments. Special assessments allow some or all of the costs associated with a public improvement to be borne by properties within a defined area benefiting from the project. An assessment is a compulsory levy made against property and can be a one-time fee or a recurring charge over a period of years. Courts in a number of states have ruled that the special assessment concept is distinct from taxation, and thus not limited by voter approval and debt restrictions. Assessments may be levied by independent special purpose governments, but often are levied by dependent areas created as "special assessment districts" by surrounding general purpose government units.

Special benefit assessment techniques are sometimes considered politically more attractive than other kinds of district and authority public works financing techniques, because theoretically, only properties directly benefiting from an improvement are assessed. Because *benefit*, rather than measurable units of a specific service, is being provided by such projects, they lend themselves somewhat better to the renovation and repair of public works than do other more traditional kinds of revenue bond financing based on user charges.

Laws have existed in many states since early in the century for special assessment districts for water, sewer improvements, local streets, some drainage projects, and so on. Water and sewer provision involves linear systems that lend themselves to simple formulas for determining assessments—frontage foot measurement, land area, and building area. Increasingly, however, special assessment financing has been extended to nonlinear systems, which involve much more complicated assessment determination. For example, the Denver city charter allows for the maintenance of a transit mall in downtown Denver to be funded through special assessments charged to property owners in the assessment district. In Los Angeles, commercial property in a special benefit assessment district is assessed at a particular rate per square foot in order to support a fixed-rail transit system.

In spite of growing interest on the part of government officials in using special assessments, there are often legal problems associated with this method of financing. Property owners may challenge either the establishment of the assessment district, or the formula used to determine the assessment, or both. Especially controversial are those cases in which some alternative method of benefit measurement indicates that the total amount of assessments—though perhaps enough to make a necessary contribution to project expenses—is actually higher than the benefits to assessed properties.

Leasing. Leasing by special purpose governments is similar to installment sales contracts used in the private sector. The technique has been increasingly used to finance the acquisition of equipment and real property without impairing the credit standing of the issuer. Typically, a special purpose government issues bonds to finance construction of a facility, and the user of the facility (often a municipality) leases it from the issuer. Lease payments are used to finance debt service payments. Often, with the repayment of the debt, the ownership of the facility passes to the user (i.e., a lease-purchase arrangement).

Lease-purchase financing mechanisms tend to be complicated, and arrangements vary from state to state. A particular advantage of such leasing is that it is considered a contingent liability rather than a general direct debt of the issuer, thus it circumvents debt limits or the need for approval of bond issues by a district's voters. At the same time, however, lease payments typically are raised through annual or biannual appropriations by the unit using the service. Thus credit ratings and investor reaction to lease-backed revenue bond issues rest heavily on the perceived commitment on the part of a legislative body or municipal government to make necessary appropriations. Lease-backed debt may or may not be rated as highly as the GO debt of the unit leasing the facilities or equipment, depending on the general features of the lease obligation.

Impact on Output Decisions

There is no general evidence that special purpose governments make better developmental decisions than other forms of community organization. The vast policy literature on transportation, water resources, and other sectors is the basic source for this topic. Most of the case studies and public policy research suggest that variables of finances, scale, technology, and political leadership have more to do with policy outcomes than does the organizational framework.

Clearly, special purpose governments that attempt to be self-financing, or that do represent small community interests, do not readily or voluntarily enter into large cooperative or regulatory relationships when significant externalities arise. Various case studies attest to this: case studies of efforts to regulate Texas and California water districts; case studies of efforts to get automobile-oriented, profitable authorities to contribute to local streets and rail transportation; case studies of attempts to get sewerage districts to upgrade wastes disposed to reduce water pollution; and case studies of airport management resistance to neighboring community demands. In effect, it is the larger, general purpose governments to which these issues fall for resolution.

This is as it should be, because all of the issues require political trade-offs. But this relationship between the implementing authority or district and the policy-setting general government needs clarification. The role of general governments in providing subsidies to special purpose governments has outgrown their role in setting policies and priorities to deal with externalities of individual districts and authorities. Generalizations from the literature indicate that even though most special purpose governments are not self-supporting, as a class they lean toward the more revenue-producing services. They have contributed more to auto-oriented transportation services than to rail transit, for example. They have contributed more to water supply and waste disposal than to treatment and pollution control. They have contributed more to major toll highways than to local access road systems.

Both supporters and opponents of the special purpose government form claim economies of scale. The argument is largely irrelevant. Larger or smaller scale can be provided by district or municipal organization, by authority or urban county organization. It is only at the intercounty, river basin, or interstate levels that the independent unit is necessary to meet scale requirements. Scale is one of the major problems with special purpose governments as they exist, however. Most of

TABLE 5.3 The Authorization and Sale of Debt Issued by State-Level Special Purpose Government Entities (SPGs)

Type of Government Entity	Authorization of SPG Debt	Structure and Sale of SPG Debt
Central Executive Branch Finance Office	3	3
Executive Branch Commission or Board	3	1
Joint Legislative/Executive Commission or Board	3	2
Legislative Majority	8	
Individual SPGs	32	44

SOURCE: Hackbart & Leigland (1990).

the examinations of independent water districts (drawing directly on natural resources) indicate serious problems of scale. Increasingly, the technical dimensions of water and waste disposal require broad policy decisions and coordination. The role of state planning and state law, of coordination by state agencies, is the single most important route to improving these relationships. The state may rely on its own action or enforce regional coordination. In either case, leadership by state government and inducement by federal government are crucial.

The need for such increased state leadership in these activities has been underscored by recent studies that detail the almost total lack of involvement in the borrowing activities of special purpose governments by officials of general purpose government (see Hackbart & Leigland, 1990). Table 5.3 indicates the numbers of state governments using various mechanisms for authorizing, as well as structuring and selling, debt issued by state-level special purpose governments (public authorities, special districts, and similar entities created by state government). The table shows that in the overwhelming number of cases, decisions about these matters are left to the special purpose governments themselves.

In only a few cases are officials of regular state government involved. For example, in Kentucky, the Office of Investment and Debt Management in the State Finance Cabinet must authorize and sell all bonds issued by state authorities. In other words, this office authorizes the specific proposal to issue debt (based on a review of state capital budgeting considerations); assembles the team of expert advisers involved in the sale (bond counsel, financial adviser, other legal advisers, accountants, auditors, and other experts); designs and structures the bond issue (and make decisions regarding the precise size, timing, nature of the debt instrument used, and whether to insure the issue); selects an underwriter or underwriting syndicate; makes arrangements for credit ratings, and so on.

Some of the advantages of this centralized approach to authorization and sale are obvious. Debt can be efficiently targeted at statewide development priorities. The costs of debt issuance can be substantially reduced by coordinating the issuance of different kinds of state debt, which might otherwise compete for space in the portfolios of large institutional investors. The state government can keep track of all debt that is considered ultimately to constitute at least contingent liabilities of state government. In cases of default, state government would be expected to repay these obligations. In some states, such centralization would be difficult because of the number of special purpose governments involved, or the size and frequency of debt issues. But some degree of central oversight seems called for. States cannot continue to rely solely on the individual directing boards of these entities to adequately coordinate development and borrowing activities, particular when so much borrowing is now needed to finance infrastructure improvement.

■ Summary and Conclusions

Although available data on special purpose governments is incomplete, it suggests a number of conclusions:

- Special purpose governments now may account for as much as 42% of all state and local government spending on public works.
- Although special purpose governments depend heavily on user charges for revenue, they tend not to use pricing methods considered to be most efficient and are not typically self-supporting. Half of all revenue comes principally from intergovernmental aid—this figure would probably be higher if revenue data were available on special purpose governments classified by the Census Bureau as "subordinate agencies."
- Special purpose governments probably issue more than 50% of all bonds used to finance infrastructure investment, though exact figures are not available.

- The ability to sell revenue bonds, and thus circumvent state and local restrictions on public borrowing, appears to be a principal reason for the proliferation of special purpose governments.

- Although access to the credit markets is widely seen as a strength of the special purpose government mechanism, successful debt financing does not guarantee project viability, may lead to relatively little funding for repair and rehabilitation, and may ultimately crowd out some other public purpose borrowing from the markets.

- Particularly with regard to borrowing, many special purpose governments appear to operate virtually without guidance, coordination, or oversight of parent government officials.

- There is no evidence that special purpose governments make better development decision than other forms of community organization.

- Tremendous variety exists within and among the various sectors of public works provision; special purpose governments range from tiny sewer districts to huge multipurpose entities that manage and finance airports, office building, mass transit, port facilities, as well as regional economic development activities.

Special purpose governments clearly play a crucial role in financing and managing public works provision. But as the above review suggests, there are serious questions about whether that role should be significantly expanded to help meet financing and management challenges posed by the nation's deteriorating public works inventory.

First, special purpose governments already spend and borrow far more for public works provision than has been recognized by proponents of an expanded role for this type of entity. Most current estimates of special purpose government spending rely exclusively on incomplete Census Bureau data. Consideration should be given to what limits realistically can and should exist regarding the large and rapidly growing role of special purpose governments in this area.

Second, given the size of the role already played by these entities, a principal weakness of the special purpose government mechanism also requires careful consideration: Many state and local government commentators and study commissions have identified a tendency for authorities, districts, commissions, and the like to remain isolated from broader policy planning frameworks. If this is the case, then many of the dollars being raised and spent by these entities in public works provision may not go to communities' most pressing needs.

Third, many of the positive characteristics of special purpose governments can be given to line agencies of general purpose government, to allow them to undertake enterprise-type activities efficiently without giving up executive and legislative controls. For example, regional or local enterprises or separate executive agencies can be created by state governments and administered by representative commissions, intergovernmental boards, or executive directors appointed by and reporting to the governor, mayor, or county executive. Fully dependent entities, such as special taxing or service areas, can be created to realize many of these advantages. All of these entities may have special powers designated by statute, without independent status. Segregated enterprise funds can be administered by executive departments. Such funds can facilitate protected financial integrity, revenue bond borrowing (if provided for by state constitution or statutes), without requiring separate administrative bureaucracies.

Fourth, various measures of state taxing, spending, and borrowing capacity indicate that many state governments could safely play much stronger roles in public works provision than they do presently. By doing so, states could alleviate some of the pressure on local governments to rely so heavily on special purpose units. For example, Texas now faces the need to rebuild a deteriorating state public works system—but the state government is not yet borrowing to finance improvements. As of 1990 Texas ranked no higher than 40th among the states on each of the principal debt per capita measures developed by Moody's Investor Service, and had $4.37 billion in state debt authorized by the state legislature, but not issued (Select Interim Committee on Capital Construction, 1989).

State laws also often unnecessarily restrict the abilities of local governments to finance infrastructure improvements without the creation of special purpose governments; eased restrictions on local government borrowing have made less necessary the creation of special purpose governments in a number of states (Bollens, 1986). States also have at their disposal a number of relatively low-cost methods for assisting local governments with debt financing for public works (see Lamb, Leigland, & Rappaport, 1993). In other words, alternatives to expanded use of special purpose governments exist. Unfortunately, most of these alternatives are far less convenient than the continued use of special purpose units. Many states need to make significant progress in their own capital planning, budgeting, and debt management practices before they can help local governments (see Hackbart & Leigland, 1990;

Yondorf & Puls, 1987). State constitutions also are often obstacles to change. Many state constitutions prohibit regular government units from enjoying the sort of legal clarity, adequate access to the bond markets, and ability to serve multiple jurisdictions that only special purpose government forms can provide. State constitutions may of course be amended, but only with much political difficulty.

Without considerably more comprehensive, comparative national data on the activities and impacts of these entities, officials at all levels of government will be unable to arrive at cost-effective policy decisions regarding their use for raising money in the capital markets, constructing and maintaining public works facilities, and providing related services. The federal government, particularly the U.S. Census Bureau, would appear to be the appropriate lead actor in the collection and analysis of such data (see Leigland, 1990).

REFERENCES

Advisory Commission on Intergovernmental Relations (ACIR). (1986). *Significant features of fiscal federalism* (1985-86 Ed.). Washington, DC: Government Printing Office.

Bollens, S. A. (1986, Fall). Examining links between state policy and the creation of local special districts. *State and Local Government Review, 3,* 117-124.

deHaven-Smith, L. (1985). Special districts: A structural approach to infrastructure finance and management. In J. C. Nicholas (Ed.), *The changing structure of infrastructure finance* (pp. 59-77). Cambridge, MA: Lincoln Institute of Land Policy.

Hackbart, M. M., & Leigland, J. (1990, Spring). State debt management policy: A national survey. *Public Budgeting & Finance, 1,* 37-54.

Lamb, R., Leigland, J., & Rappaport, S. (1993). *The handbook of municipal bonds and public finance.* New York: New York Institute of Finance.

Leigland, J. (1990, Spring). The census bureau's role in research on special districts: A critique. *Western Political Quarterly, 3,* 367-380.

Leigland, J. (1993). Overview of public authorities and special districts. In R. Lamb, J. Leigland, & S. Rappaport (Eds.). *The handbook of municipal bonds and public finance.* New York: New York Institute of Finance.

Leigland, J., & Lamb, R. (1986). *WPP$$: Who is to blame for the WPPSS disaster.* Cambridge, MA: Ballinger/Harper & Row.

National Council on Public Works Improvement (NCPWI). (1986). *The nation's public works: Defining the issues.* Washington, DC: Author.

National Council on Public Works Improvement (NCPWI). (1988). *Fragile foundations: A report on America's public works.* Washington, DC: Author.

Peters, S., Lovette, S. S., & Choe, K. T. (1993). Emerging borrowing priorities—Financing infrastructure. In R. Lamb, J. Leigland, & S. Rappaport (Eds.), *The handbook of municipal bonds and public finance* (pp. 788-806). New York: New York Institute of Finance.

Petersen, J. E., Holstein, C., & Weiss, B. (1993). The future of infrastructure needs and financing. In R. Lamb, J. Leigland, & S. Rappaport (Eds.), *The handbook of municipal bonds and public finance* (pp. 724-769). New York: New York Institute of Finance.

Peterson, G. E., & Miller, M. J. (1982). *Financing urban infrastructure: Policy options.* Washington, DC: Urban Institute.

Porter, D. (1987, May). Financing infrastructure with special districts. *Urban Land, 46,* 9-13.

Porter, D., Lin, B. C., & Peiser, R. B. (1987). *Special districts: A useful technique for financing infrastructure.* Washington, DC: Urban Land Institute.

Select Interim Committee on Capital Construction. (1989). *Interim report to the 71st Texas legislature.* Austin: Texas Legislature.

U.S. Department of Transportation, Federal Highway Administration. (1991). *Status of the nation's highways and bridges: Conditions and performance, 1991.* Washington, DC: Government Printing Office.

Vaughan, R. J. (1983). *Financing public works in the 1980s.* Washington, DC: Council of State Planning Agencies.

Walsh, A. H. (1978). *The public's business: The politics and practices of government corporations.* Cambridge: MIT Press.

Yondorf, B., & Puls, B. (1987). *Capital budgeting and finance: The legislative role.* Denver: National Conference of State Legislatures.

6 Public Works and Public Dollars: Federal Infrastructure Aid and Local Investment Policy

HEYWOOD T. SANDERS

Over the past decade—from *America in Ruins* of 1981, to *Fragile Foundations* of 1988, to the *Financing the Future* report of the Infrastructure Investment Commission in 1993—studies decrying the nations "decaying infrastructure" have become a staple of domestic policy debates (see Choate & Walter, 1981; NCPWI, 1988). In general, these studies have urged a reordering of national priorities and a massive increase in infrastructure investment. The widely read *Fragile Foundations* report, for example, recommends doubling current infrastructure spending, with total outlays reaching $2-3 trillion by the end of the century.

Frequent reports of a looming infrastructure crisis, the public investment proposals of President Clinton, and concern about the condition of the national economy have kept infrastructure at the forefront of economic policy discussions. Yet despite more than a decade of national attention to underinvestment and structural deficiency, policy analysis and recommendations in the popular press remain consistently dispiriting. Unless we double or triple infrastructure spending, says the conventional wisdom, we will forfeit future economic growth and face growing bills for repair and renewal.

In an alternative approach to summaries of enormous national needs, this analysis seeks to explain how we invest in infrastructure and why streets and sewers in many parts of the country appear to be failing. It goes on to suggest how infrastructure investment reflects a process of political choice at the state and local levels. Finally, it provides some future directions for national infrastructure policy that seek to alter and define the incentives to state and local governments. For example, federal policy should structure incentives, enabling the complex federal-state-local partnership to deal more effectively with the problems of decay and deferred repair. Yet without an understanding of how infrastructure decisions are made, a new federal effort to boost spending may well prove counterproductive, or of limited import.

■ The Status of the Nation's Infrastructure

To a great extent, the notion that the nation faces an "infrastructure crisis" rests on a small number of themes, regularly repeated in a host of studies. First, many studies argue that, relative to the 1960s, the nation is underspending on public facilities. Aggregate investment trends, they argue, indicate a worrisome neglect of fundamental national needs. Second, these studies use national statistics on highway and bridge conditions to demonstrate substantial (and growing) deficiencies in structural quality and, presumably, safety. These arguments have generated concern both about whether we are spending enough today and about growing investment requirements in the future.

A broader historical view suggests, however, that contemporary spending levels should not cause alarm. Figures on state and local infrastructure investment indicate that spending has risen sharply in recent years, particularly for highways. Indeed, current capital spending approximates (in inflation-adjusted terms) the peak years of this century. Statistics on highways and bridges tell a parallel story. What problems that do exist today are often concentrated in a remarkably few states and locales. Moreover, highway statistics on system performance and pavement quality generally indicate that the nation's roads are getting better. Although these themes have not been part of the debate on national infrastructure policy, they are critical both to understanding our contemporary circumstances and to developing effective national policy.

Spending Levels

Infrastructure deficiency and decay are often attributed to inadequate public spending, particularly in comparison with the spending levels of the 1960s. *America in Ruins* (Choate & Walter, 1981), for example, notes a sharp decline in capital spending from 1965 to 1977, the last year included in the book's analysis. A more recent study, *Fragile Foundations* (NCPWI, 1988), tracks aggregate spending from 1960 to 1985, concluding that the numbers are discouraging, and suggesting a return to the investment levels of previous decades.

Aggregate spending is not, however, appropriate for evaluating infrastructure policy.[1] The fact that total spending on infrastructure was a greater proportion of gross domestic product in the past than at present tells us absolutely nothing about *how much* we should spend today to sustain the performance of the economy. Nor do such aggregate numbers tell us *what kinds* of investments would most benefit the economy— should we invest in new construction or maintenance, wastewater treatment or aviation facilities? Indeed, the more substantive issue is what public goods we are buying with our infrastructure investment dollars, and how those goods will aid the nation and its economy in the future.[2]

Investment Cycles

Even a limited analysis of investment over time yields some sense of what propels public spending and public priorities. Most important, perhaps, is the fact that *capital spending on infrastructure has historically occurred in a series of broad—and varied—cycles that can make individual-year comparisons highly misleading.* Figure 6.1 outlines these postwar cycles, expressed in terms of constant 1987 dollar spending on basic infrastructure.

A comparison of spending in the late 1970s or early 1980s with that of the previous decade presents a dismal image, with investment declining steadily, presumably because the nation placed a lower priority on public capital investment. Most reports warning of the looming "infrastructure crisis" present this picture. By contrast, a comparison of total capital spending in 1991 and 1968 (the peak year for postwar infrastructure investment) demonstrates roughly equivalent annual investment.[3] This news often gets overlooked in discussions of infrastructure. Some analyses of infrastructure employ a broader definition of infrastructure, including railroads and airports. The Appendix (Figure A.1) indicates that spending based on a broader definition of infrastructure capital mirrors the pattern noted here.

Even more revealing is an examination of spending trends for individual infrastructure categories. Different types of infrastructure, or "modes," have historically followed quite distinct cycles. As Figure 6.2 illustrates, state and local spending (including federal grant aid) on

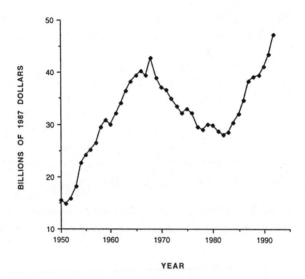

Figure 6.1. State and Local Basic Infrastructure Spending by Year
NOTE: Basic infrastructure includes highways, streets, bridges, and tunnels; sewers and sewage treatment plants; water distribution and supply systems.
SOURCE: Data from the Bureau of Economic Analysis, U.S. Department of Commerce (1993).

sewer systems and wastewater treatment plants rose in the 1970s, even as spending on highways and bridges slowed. The reverse was true in the 1980s: Spending on sewer systems and wastewater treatment plants fell, and highway and bridge spending accelerated. *As these sharp variations in spending patterns indicate, all types of public capital have not received the same budget treatment or public-sector priority in recent decades.*

Although several factors contribute to the periodic rise and fall of the public investment cycle, two stand out—broad economic and political trends, and federal policy.

Broad National Trends

Broad economic and political trends affect the volume of resources invested in infrastructure. Investment varies with economic conditions and political preferences. The one predictable thing about current infrastructure spending is that it too will change. Policy responds to changing public demands and preferences—in a word, to changing priorities.

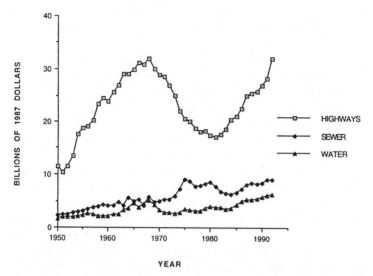

Figure 6.2. State and Local Capital Spending for Highways, Water, and Sewer Systems by Year
SOURCE: Data from the Bureau of Economic Analysis, U.S. Department of Commerce (1993).

Highway and bridge spending, for example, illustrates the forces that affect public investment.[4] Figure 6.2 shows that road spending began a steep upswing in the early 1950s. This surge reflected a growing demand for new roads to keep up with housing development, as well as a need to redress the investment drought of the Depression and war years. Ultimately, however, this increase was a product of a new federal program, the interstate highway system. Yet even that program was sold in part—as the title, National Defense and Interstate Highway Program, indicates—for its critical role in moving defense materials in the Cold War era.

Highway investment grew as long as the economy prospered and the public was receptive to increased construction. But during the economic stresses of the 1970s and early 1980s, real highway spending declined. Inflation eroded the purchasing power of a fixed gasoline tax, and high interest rates slowed the issuance of new state and local bonds. In addition, a dawning environmental awareness spurred a "freeway revolt" in places like Seattle, San Francisco, Baltimore, and the District of Columbia. This movement slowed highway spending in the 1970s by encouraging both limits on construction and transfers of resources to mass transit projects. The result was a clear increase in capital spending on mass transit systems, buoyed by increased federal grant aid. In recent years, highway spending has begun to return to the annual levels of the 1960s, prodded by greater public concern about the infrastructure and by an increasing acceptance of higher taxes and fees for highway purposes.

State and local government highway investment in 1992 is estimated at $29 billion (in constant 1987 dollars). That figure represents a sharp increase from the $26-billion spending level of 1991, and a dramatic boost of more than 67% from the $17.3 billion in highway capital spending in 1980. This spending boost is a combined product of increasing state spending, buoyed by state gasoline tax increases in the past few years, and a greater volume of federal grant funds for highways. And the passage of the Intermodal Surface Transportation Efficiency Act in late 1991 promises even greater federal support for highway investment. Indeed, even in the absence of new investment initiatives by the Clinton administration, federal highway aid is projected to rise from $14 billion in fiscal 1991 to more than $18 billion in 1995. The Clinton administration investment proposals call for an additional $5.6 billion in new outlays over the 1994 to 1997 period. And these increases in federal highway spending will likely spur additional state-level capital spending.

The large-scale change in highway investment since the early 1980s suggests that governments, and the public, are indeed capable of supporting infrastructure investment through increased taxes and fees. Yet that political response is not consistent either across the states or across different types of governments.

Gaps in the Image of Nationwide Decay

Many studies use Federal Highway Administration (FHWA) statistics on highway and bridge conditions to support the case for a national infrastructure crisis. These data, it is said, reveal that American roads and bridges are crumbling—even approaching collapse. According to this view, the national transportation system will not be adequate until it is free from all deficiencies. Studies regularly cite the FHWA's figures because they come from a detailed national database that provides the most credible and comparable basis for assessing infrastructure conditions.

The FHWA estimates that the full price tag for remedying the nation's deficient and obsolete bridges will come to about $165 billion over the next 20 years, requiring annual spending of about $8.2 billion. Current federal estimates put the cost of today's backlog of bridge deficiencies at a total of $78 billion. These estimates of bridge needs include a wide range of deficiencies across all of the states. They also include both bridges on the federal-aid highway system and those that have historically been the responsibility of state and local governments. As bridge problems are disaggregated, in terms of differing categories of need and sharply varying state-level conditions, a picture of far more modest and reasonable requirements emerges.

Degrees of Deficiency

The FHWA classifies bridges as deficient for many different reasons, some of which are much less important than others. Some bridge problems, for example, do not reflect safety or structural conditions, but pertain to traffic capacity or design features. This latter grouping of deficiencies is termed "functionally obsolete." Although these spans may not meet contemporary design standards, they can often be made adequate simply through changing signs, the nature of the highway approach, or the roadway striping. Such modest improvements, far less expensive than actual bridge rehabilitation, allow for greater safety and more efficient traffic movement. Indeed, about half the $165-billion sum of bridge "needs" is represented by these "obsolete" structures.

Of most concern are bridges with real safety problems or structural inadequacies. These are termed "structurally deficient" by FHWA, and commonly fail to meet contemporary standards for bridge load-carrying capacity. It should be noted that even some newly built spans are classified as "deficient," because state or local governments have chosen to build them with a lesser load capacity than federal standards. Thus even "deficient" bridges are not necessarily safety problems; they simply meet earlier and lesser standards for loads.

Structurally deficient bridges account for slightly more than half the aggregate cost of bridge renewal or replacement. By focusing investment on the bridge structures that most warrant attention, the annual cost of bridge renewal over the next 20 years—about $5 billion—is equal to what we are currently spending on bridge renewal.

Even improving or replacing these structurally deficient spans may not necessarily require the total investment estimated in the $165-billion sum. Some low-cost improvements, or reduced weight limits, might suffice to maintain the serviceability of many of these structures. Thus the General Accounting Office has written that "while 136,000 bridges classified as structurally deficient may have deteriorated to the extent that they cannot carry the load for which they were designed, with proper load-posting they can safely serve existing traffic" (U.S. General Accounting Office [GAO], 1988).

Distribution of Deficiencies

Although the popular perception is of a truly national crisis, with problems distributed roughly equally across the nation, FHWA bridge statistics for mid-1992 reveal a concentration of structural deficiencies in a relatively few states. This pattern is not the result of weather or culture, for individual states can often boast, or despair, of bridge systems in conspicuously different condition from those of neighboring states.

For the nation as a whole, about 12% of bridges on the federal-aid system were classified as structurally deficient in the most recent FHWA tally—a total of 33,800 structures (GAO, 1988). Yet despite consistent federal policies across the nation, some 4,609 deficient bridges, or about 13% of the national total, are found in just one state, New York. Three other states—Missouri, Pennsylvania, and Ohio—account for another 18%. Thus, just four states (with about 15% of all the federal system spans) possess almost one third of the nation's structurally deficient bridges.[5]

If we look at deficient bridges within individual states, we can see an equally striking pattern. A small number of states appear to have a serious problem with bridges. New York State leads the list, with more than 51% of its federal-aid spans deficient. The second- and third-worst states, West Virginia and Missouri, are about tied at 23% structurally deficient. Yet half the states in the union have 11% or fewer in deficient condition.

Simple factors such as region or weather do not explain the variation in bridge problems. New York at 51% deficient can be compared to neighboring New Jersey (23%) or Connecticut (14%). West Virginia (23.7% deficient) can be compared to Virginia (10%) or Kentucky (5.1%). Elsewhere, Illinois (13%) appears markedly better than Wisconsin (21% deficient). Indeed, in the upper Midwest, Wisconsin has about twice the incidence of deficient structures as Minnesota (10.8%) and Iowa (9.5%).

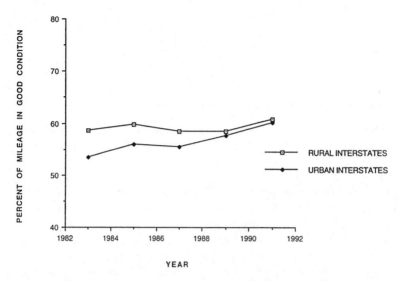

Figure 6.3. Percentage of Interstate Highway Mileage in Good Condition by Year

SOURCE: Data from the Federal Highway Administration, U.S. Department of Transportation (1992).

Highway Conditions and Variations

The nation's major highways repeat the pattern of national bridge conditions. First, overall national roadway quality, measured by the FHWA's Present Serviceability Rating (PSR) scale, was stable or improving from 1983 through 1991, as shown in Figure 6.3 (U.S. Department of Transportation [DOT], 1993).

These national PSR figures do not support the notion that the nation's highways are falling apart. Moreover, the same measures of road condition indicate substantial variation across the states, just as with bridges. Although figures on interstate highways from 1991 (the most recent) show less than 8% of the total mileage in poor condition, a handful of states show a strikingly high incidence of poor pavement. More than 38% of New Mexico's interstate mileage is rated deficient, with about 37% in Missouri and 30% in Colorado. At the other end of the spectrum, Alabama, Delaware, and Wisconsin had no poor-rated urban interstate miles. As with bridges, notable differences occur among neighboring states, thus undercutting explanations based on weather, highway environment, or political culture. Missouri's 37% is far different from the

4% of Illinois, or the 5% in Kansas and 2% in Indiana. And some states have even managed a striking turnaround. In the South, Mississippi— long one of the worst states, with 30% of its interstate mileage in poor condition in 1985—now shows no interstates in poor condition.

A Revised Image of the Nation's Infrastructure

This brief review of spending cycles and highway conditions suggests two conclusions. *First, actual needs are far more limited and concentrated than is often supposed.* Serious bridge condition problems are highly concentrated in a small number of states, rather than being pervasive across the country. And highway conditions follow a parallel pattern, with a relatively small number of states demonstrating unusually serious problems. This pattern of variation and concentration suggests that many state and local governments have succeeded in recognizing their infrastructure needs, and in providing for them.

Second, the skewed distribution of needs reflects state investment and management choices, rather than a pervasive failure of federal policy or public concern. No other explanation can satisfactorily account for an acute bridge or highway problem in one state, and only mild deficiencies in a neighboring state. Despite the argument that infrastructure is "invisible" or of little concern to the public, some states clearly have made infrastructure renewal and investment a major priority. Ohio, for example, has relatively low proportions of both deficient bridges and highways in poor condition. Iowa, too, has a record of high-quality roads and highways, and a history of sustained public investment. And Mississippi has managed a dramatic improvement in its pavement conditions in recent years. In responding to the nation's current infrastructure problems, federal policy must recognize that individual states have historically performed quite differently and have set varying budget priorities: These differences are ultimately the product of politics and political values.

■ Sources of Infrastructure Problems

The Role of State and Local Political Choice

Although studies of infrastructure regularly document total national needs and a host of individual "horror stories," they rarely examine how the investment preferences and choices of state and local governments have contributed to the current state of our infrastructure. State and local officials operate in a political environment, where they must constantly balance competing interests and electoral necessities. They also face both the constraints of federal grant requirements and the opportunities of federal dollars, which are not tied directly to their own revenue raising. Choices of infrastructure investment are thus products of a complex balancing act, even in the face of federal aid and direction. Simply put, state and local decisions "filter" infrastructure outcomes. That filter effect goes far in explaining why some elements of the infrastructure are receiving less attention than others.

Investing in New Construction

Although the current rhetoric of infrastructure need invariably stresses aging facilities and potential safety problems, state highway capital investment has continued to emphasize new construction and the widening of existing highways to add capacity. Figure 6.4 shows that new construction consumed 26.8% of state-funded highway capital investment in 1981. And although that figure dropped over the early years of the decade, it has since risen to 23.6%. About another 10% of state spending each year supports projects classified as major widening.[6]

The most recent FHWA figures on capital spending are not fully comparable with earlier numbers. They nonetheless substantiate the new construction emphasis. In 1991 some 42% of state highway capital spending was devoted to "system preservation"—improvements to existing roads and bridges. But 51% was spent on "capacity additions," including both major widening projects and entirely new roads and bridges (20% of the total). Thus, projects that add new roads or additional capacity still make up the majority of state capital spending.

The 1991 Surface Transportation Act's new focus on the linkage between highways and clean air may begin to alter the traditional bias in favor of new construction, particularly in metropolitan areas with serious air quality programs. And the 1991 Act's provision of increased local flexibility in choosing among highway and mass transit projects may also reduce the new highway development bias. Yet it is equally likely that state and local officials will continue to promote projects like downtown people-movers and light rail systems, as well as new highways, that hold out the promise of economic development.

The central place of new construction may not fit the image of widespread infrastructure decay, but it makes a great deal of political sense. New construction projects carry substantial rewards for state and

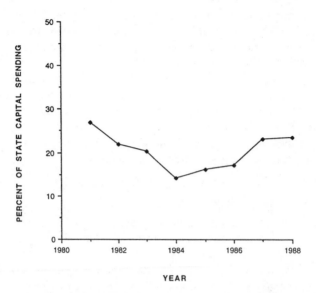

Figure 6.4. Percentage of State Highway Capital Spending on New Construction by Year

SOURCE: Data from the Federal Highway Administration, U.S. Department of Transportation (1990).

local elected officials: They can be used to attract new jobs and industry, to relieve congestion in developed areas, and to promote new urban development and suburban expansion. And just as new construction spurs physical development, it also promises increasing property tax wealth and the expansion of government revenues.

The political preference for spending on new construction evident in state highway investment is paralleled across other modes of infrastructure. New water and sewer lines translate into new residential and commercial development and an ever-greater property tax base. In contrast, the replacement of an aging water pipe or the remedying of a neighborhood flooding problem may offer only a limited political and fiscal return.

Economic Development Bias

State and local governments commonly see themselves in a contest to attract and subsidize new jobs and industries. That contest takes form in a variety of public policies, from tax advantages such as property tax abatements and inventory tax exemptions, to direct public assistance through new infrastructure or tax-exempt bonds. The search for economic development reflects some realities of state and local fiscal needs, although it can be quite wasteful of public resources. More important though, it is also good politics, enabling mayors and governors to garner the credit for job creation and economic expansion in the short term.

The political rewards of economic development lead governments to choose spending projects that promote presumed job growth, rather than those that merely sustain or renew existing infrastructure. This element of choice by city governments is illustrated by the changing spending emphases under the Community Development Block Grant (CDBG) program.

Begun by the federal government in 1974, the CDBG program offered block grant funds for city governments that could be used for a range of public works and neighborhood improvements. The relative freedom to invest in everything from basic infrastructure to housing rehabilitation should demonstrate, in the aggregate, the priorities of local governments.

In the early 1970s the city governments receiving entitlement funds concentrated on traditional urban renewal and land clearance activities. They also devoted a substantial portion of their federal-aid dollars to public works—street improvements, sewers, bridges, and other basic infrastructure. Yet through the 1980s the proportion of CDBG funds devoted to public works and improvements declined, as shown in Figure 6.5 (U.S. Department of Housing and Urban Development [HUD], 1992).

For 1991, public works and facilities accounted for only 19% of CDBG spending. Then, just as spending on traditional infrastructure declined relatively, spending for economic development grew. The economic development category amounted to just 3.4% of spending in 1979; it reached 12.6% in 1988. Planned economic development spending has declined in the past 3 years as a result of tighter federal restrictions, so that in fiscal 1991 it stood at 8.1% of CDBG entitlement spending. Given a relatively free choice, using federal aid for economic development is terribly appealing to state and local officials.

Economic Development Infrastructure

Investment support for economic development doesn't stop with federal community development aid. Increasingly, state and city governments are directing their political and financial resources toward investment in such things as sports stadiums, convention centers, arts

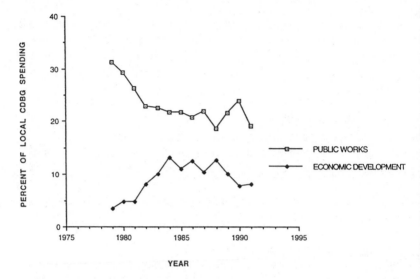

Figure 6.5. Local Community Development Block Grant Spending by Purpose and Year
SOURCE: Data from U.S. Department of Housing and Urban Development (1992).

and entertainment complexes, aquariums, and other tourism-related facilities. We can term these projects *economic development infrastructure,* aimed at the promotion of the local economy.

The emergence of economic development as a significant object of spending, relative to basic infrastructure, reflects a change in political priorities and fiscal direction. It is not that fiscally strapped central cities like Philadelphia or St. Louis have cut back on capital spending for infrastructure in recent years, but that they have used public dollars to support a whole set of new spending projects. Although these new investments may well help maintain the economies of these central cities, they compete for political and financial capital with traditional public works, which are nonetheless portrayed as an acute need.

Convention Centers

Data on the size of convention facilities in America's major cities reflect this concern with development infrastructure. In 1969 these centers included 6.6-million square feet of exhibit space.[7] By 1980 the total had reached more than 11-million square feet. The 1990 figure is

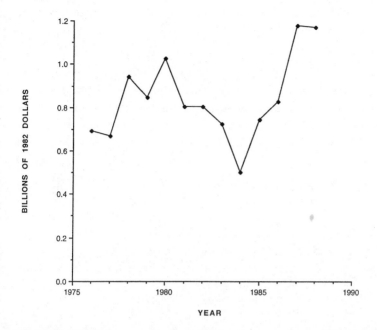

Figure 6.6. State and Local Capital Spending for Amusement and Recreational Buildings

SOURCE: Data from the Construction Division, Bureau of the Census, U.S. Department of Commerce (1990).

almost 17.2 million—an increase of more than two-and-a-half-times over two decades.

This investment in convention and exhibition facilities is costly, at a time when local governments bemoan their fiscal problems. The Census Bureau has tracked figures on state and local capital spending for amusement and recreational buildings (a category that includes convention facilities) since 1976.[8]

As Figure 6.6 shows, this construction spending amounted to $691 million in inflation-adjusted dollars in 1976. Annual spending came to $804 million in 1982, and $1.2 billion in 1987. The most recent figure, for 1989, repeats the $1.2-billion annual total, the highest spending rate since the series began. Moreover, the public investment in convention facilities has not slowed. Big cities from San Francisco to Providence will be adding millions more square feet over the next few years. And in places like Philadelphia, Denver, and St. Louis, they will be aided by state governments as well.

Data on long-term bonds also point to a shift in local priorities.[9] New bond issues by state and local governments for roads, highways, and streets totaled $5.56 billion in 1992. While state and local governments were generating capital funds for highway investment, however, they were also issuing bonds for a wide range of development infrastructure. Convention center bonds came to $1.4 billion for 1992, with another $560 million for stadiums, and $2.6 billion for other economic development purposes. That comes to a grand total of $4.598 billion for economic development—a sum not much smaller than that for streets, highways, and bridges. *Thus, despite the consensus that the need for public infrastructure investment is dire, local officials have concentrated much of their energy and resources on economic development.*

The Case of Cincinnati

The investment choices of a single city government illustrate the transformation in local capital spending. An overview of capital spending in Cincinnati, Ohio, is shown in Figure 6.7 at 5-year intervals.

In parallel with other governments, Cincinnati devoted much of its resources to street and highway spending in the 1950s and early 1960s. Yet gradually it shifted to capital spending for an array of economic development purposes, largely in the downtown area. These included urban renewal, a convention center, and the continuing subsidy of the city's stadium. Through the 1980s the city spent about one third of its annual capital budget on economic development. Recent spending has supported the downtown pedestrian skywalk system, convention center expansion, new street and park improvements, and stadium subsidy.

As long as state and local officials face their own peculiar political and fiscal pressures, federal aid and federal policy must necessarily be filtered, producing results that may be quite different from what was intended. The particular conundrum of today's infrastructure needs is that they have not occurred overnight or in a vacuum. State and local governments do face resource constraints, and many have recently raised taxes and fees relevant to infrastructure, like the gasoline tax. But they have also pursued capital spending policies aimed at things like new highway construction and convention center development, which represent important political goals. And that implies that federal policy should affect the larger issue of investment *choice*, not just investment *dollars*.

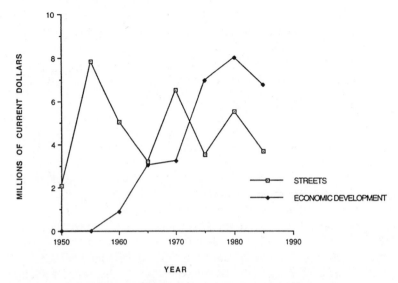

Figure 6.7. Cincinnati Capital Spending by Purpose and Year

NOTE: "Streets" include expressways, streets, bridges, and traffic engineering. "Economic Development" includes urban redevelopment, downtown improvements, downtown parks facilities, and street improvements dedicated to commercial and industrial areas.
SOURCE: Data for 1950-1975 from Cincinnati *Annual Financial Report*, selected years. For 1980 and 1985, data were coded from the Cincinnati *Capital Improvement Program* for the indicated years.

Responding to New Federal Dollars

Federal grant dollars enter a complex fiscal environment at the state and local government level. These entities must necessarily mediate competing interests and demands, and face simultaneous pressures for new spending and stable or lower taxes. As a result, federal-aid dollars often emerge, in terms of new spending by states and cities, in ways quite different from the original intention. Federal programs can offer incentives for other governmental units to shift or reduce their own spending. Or these programs can create a set of perverse incentives that bias actual results. Thus, the process of state and local decision making shapes total spending at the same time that it affects the products of these governments.

Substitution

The problem of substitution or "grant displacement" is well documented in the academic literature (see, e.g., Gramlich, 1977). Substitution occurs when federal grant dollars *replace* state and local spending, rather than *supplementing* it. In the most absolute case of substitution, a one-dollar increase in federal aid will lead to *no new spending at all*, as other units of government simply reduce their own revenue raising and spending in that particular area.

There is a considerable political rationale for substitution behavior, even though it is rarely absolute. New federal dollars offer the opportunity to produce desired goods without the need to increase local taxes. Federal aid can also provide capital funds without the political costs of local revenue raising, such as a referendum on a state or local bond issue. No voters have to be directly persuaded, and the very receipt of federal grant funds can enhance the image of a mayor or governor. Thus much of the appeal of federal aid for urban redevelopment or revitalization projects, like the Urban Development Action Grant program, lay in the announcement of a successful project and evidence of federal support.

One short-term federal-aid program for public construction, the Local Public Works program (LPW) of the 1970s, is a vivid—if extreme—example of the fiscal and political character of substitution at the local level. The LPW program distributed some $6 billion in federal grants for public construction in the 1977 to 1979 period. These funds should have spurred a substantial increase in capital spending on infrastructure. Yet as Figure 6.1 demonstrates, aggregate investment showed only a tiny increase in 1978, the peak year for federal LPW aid.

State and local governments used their new federal dollars primarily to replace their own spending. According to one analysis, "As soon as the program was enacted and its mode of implementation became apparent, state-local construction spending plummeted—presumably in anticipation of the possibility that LPW funding would be available to finance projects that otherwise would have gone forward with state-local funds" (U.S. Department of Treasury, 1985). Rather than adding to national spending on public works in total, federal LPW dollars fell victim to substitution.

These new federal funds under LPW were also affected by the filter of local political choice. Only about 39% of the aid funds were spent on basic infrastructure (U.S. Department of Commerce, 1978). Much of the balance went to develop general government buildings, cultural and recreation facilities, and economic development projects. And the availability of federal funds not subject to traditional political constraints on local spending encouraged local "pet projects." Cleveland, despite worsening fiscal conditions, invested in the renewal of its downtown Public Square and a downtown theater district. St. Louis added a pedestrian walkway connecting its new convention center to other downtown buildings. And Detroit funded the renovation of Tiger Stadium, to keep the local baseball team from moving.

As a short-term response to a national recession, the LPW program was unlikely to cause state and local governments to reshape their spending preferences. But long-term infrastructure aid programs face a parallel situation in terms of aggregate spending, where new federal dollars largely substitute for those of other levels of government. This effect is quite likely where state and local spending is a regular and continuing activity. Take the case of federal aid for wastewater treatment under the Clean Water Act. In 1972 Congress imposed stringent standards for sewage treatment and provided substantial federal aid, covering 75% of the cost, for developing better sewage treatment plants. But while total spending for wastewater treatment rose, local investment actually fell over time. Even in areas that required larger treatment plants simply to meet the pressures of urban growth and expansion, local officials recognized that federal aid would meet most of their needs. Federal-aid money substituted for local dollars.

When federal-aid policy changed in 1982, however, direct grant aid began to be phased out. By 1987 the program was shifted from grant assistance to loan aid, through revolving loan funds at the state level. Federal aid clearly dropped over this period, as shown in Figure 6.8. But state and local own-source spending also rose. Local officials had obviously had the fiscal capacity to finance sewer and wastewater treatment investments. The ready availability of federal-aid dollars had allowed them to shift the cost to the national government.

Substitution effects can be seen throughout the range of federal-aid programs. Its impact, however, varies with the constraints on federal grant dollars and the mechanisms for financing the state and local spending share. State highway investment, for example, is funded largely through dedicated taxes and fees, like the gasoline tax. A boost in federal aid would probably not induce states to reduce their own gasoline taxes, as we would expect of dollar-for-dollar substitution.

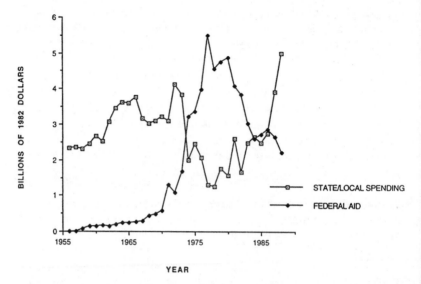

Figure 6.8. State and Local Capital Spending for Wastewater Treatment and Federal Grants

SOURCE: Data from U.S. Congress, Congressional Budget Office (1988).

Rather, state governments are likely to use new federal aid to pursue their own spending preferences. The result may be greater spending, but not on the projects or programs that are the targets of federal grant aid.

Studies of federal aid and state highway spending support just this conclusion. A 1985 analysis of federal and state capital spending on primary, secondary, and urban aid systems concluded that "each additional Federal dollar reduces State spending from discretionary funds by 62 cents" (Meyers, 1985, p. 16). But while federal dollars substituted for state dollars on the federal-aid system, these funds in turn allowed states to spend more on the unaided, or local parts of their highway systems. New grant money was spent on highways, but it allowed states to fund their own, preferred local projects. This finding mirrors economist Edward Gramlich's conclusion about federal policy in 1977: "If it wants to encourage somewhat more spending on such areas as highways where states have long been spending large amounts, its grant money will probably get lost in the shuffle anyway" (Gramlich, 1977, p. 233).

The Capital Bias

Federal infrastructure aid provides funds for *capital investment*, although some highway programs support rehabilitation and resurfacing that really amount to maintenance. Such an emphasis makes sense in a case like the Interstate highway program, where the intention was to secure participation by all the states in building an entirely new national transportation system. The problem is that maintenance remains entirely a state and local fiscal responsibility, one that is easily deferred or reduced in times of political or fiscal stress. Indeed, the attractiveness of substantial federal aid can even encourage the deferral of maintenance spending on some needs, *until structures become eligible for capital grants for replacement or renewal* (Hornbeck, 1990).

State and local governments have steadily increased their total spending on highway maintenance in recent years, and there is no evidence that their stewardship of bridges and highways in total has been improper. Nevertheless, federal grant assistance, notably in the case of bridges, does offer an incentive to limit expensive spending for repairs in favor of federal aid.

New York City provides an example of this type of result. The national Highway Bridge Replacement and Rehabilitation Program provides federal funds for up to 80% of the cost of bridge renewal. Thus, in 1988, City officials could reassure their citizens that although the rehabilitation of the Williamsburg Bridge would cost about $350 million, the federal government would pay most of the bill, with additional state aid likely. The City's ultimate share of the cost of inattention to aging and maintenance would thus be only a fraction of the true cost of repair—it faced little pressure or fiscal incentive to invest its own funds in maintenance and repair over the years.

Policy Implications

Our current infrastructure problems reflect the interaction of federal grant policies and state and local investment choices. New federal dollars may well add little to aggregate spending and be lost in the shuffle of current state and local priorities, as Gramlich concludes. Or they may support projects that better fit state or local preferences and political rewards. Finally, federal policies may create their own perverse incentives, which bias the choices of other governmental units. Such conclusions do not call for an abandonment of the federal grant assistance that has successfully built highways and improved water quality. They do suggest that federal aid should be targeted to specific problems, require a maintenance of fiscal effort by recipients, and seek to avoid the unintended consequences of fiscal bias.

■ State and Local Financing

Tapping the Fiscal Cracks

The first part of this chapter focused on the cycles of public infrastructure investment and the wide variation in the incidence of problems. A need particular to one time or place may have no relevance in another. Infrastructure problems and circumstances are simply not consistent across the nation. The same is true of state and local financing and investment. Some places, choosing to invest in basic infrastructure, have imposed the necessary taxes or user fees. Others have not. The sentiment of the voters seems quite consistent. The public has recently been quite willing to support higher spending where they know it will support infrastructure investment.

The Gas Tax

The gasoline tax has long served as a certain money generator for highway purposes. Indeed, in 1987 gasoline taxes and other user fees accounted for about 85% of total highway revenues. But in the decade of the 1970s, the fixed gasoline tax provided less and less real revenue, as inflation eroded its purchasing power. To keep up with the impact of higher-mileage automobiles and reduced purchasing power, government had to raise tax rates. In some states there was a reluctance to boost taxes, and political initiatives were often constrained by public opposition to taxation and spending generally. Still, *some states managed far better than others to keep up with the pressures of inflation and increasing highway needs.*

Iowa and California make a good comparison. Both states imposed 7-cent-per-gallon gasoline taxes in 1966, a level just about the national average. Yet by 1987 Iowa had managed to raise that tax to 16 cents. California's tax grew to only 9 cents, where it remained into 1990. And Iowa was not the only state to bear the political and fiscal cost of boosting the gas tax. Although Maryland's 1966 tax rate was the same as Iowa's and California's, by 1987 it had been increased to 18.5 cents.

Clearly, states have varied over the years in their commitment to infrastructure investment. That commitment shows some striking parallels with the data on bridge conditions, for example. New York, with the worst bridge conditions in the nation, today imposes a gasoline tax of only 8 cents per gallon—the lowest level among the 50 states. If the state had merely kept up with its inflation-adjusted taxation level of 1960 (6 cents), it would today be imposing a tax of 29 cents—almost five times higher than its current levy. New York officials properly argue that the state has historically depended on bond issues, rather than only on a dedicated tax, for highway and transportation improvements. Still New York's motorists have been paying far less in the form of gasoline taxes than the citizens of other states.

For most states, the dedication of the gasoline tax to roads and highways provides a high degree of political insulation and support. The voters can see improvements and the products of revenue generation. And they tend to support such spending, whether in the form of tax increases, or in the form of bond issue proposals. If political leaders are unwilling to propose tax increases or new bond issues, however, the public lacks even the opportunity to endorse (or perhaps oppose) infrastructure spending and investment.

Bond Financing

In some states, like New York, and most local governments, bond issues have been the backbone of infrastructure capital investment. Bond issues, though, depend on a partnership between the public officials who must propose and support them and the voters who must endorse them. Spending works only when political initiative is coupled with public support.

In the 1950s and 1960s, state and local governments relied consistently on new bond issues for infrastructure investment. In the 1970s, however, public opposition to new construction proposals and new taxes reduced the support for bonds. The result was a clear decline in the volume of bond proposals on the ballot, as city and state officials increasingly shied away from developing investment programs that required voter sanction.

Recently, however, voters have demonstrated a substantial level of support for new bond proposals. For 1989, 77.4% of the dollar volume

Figure 6.9. Volume of Nonschool Bond Proposals Voted on by Year

SOURCE: Original data from the *Bond Buyer Municipal Statbook* (1986) and unpublished Bond Buyer statistics, adjusted by a public construction price deflator.

of voted bond issues were passed, with the approval rate for transportation issues topping 84% ("Voters Gave Nod," 1990). Figures for 1992 show that 60.8% of the dollar volume of proposed bond issues were approved by the voters. And once again, voter support for transportation investment was notable, with an approval rate of 80.5%. Nonetheless, *state and local governments have often refrained from putting proposals on the ballot, avoiding the opportunity to generate new funds for public infrastructure.*

Data on the volume of nonschool bond issues over the past four decades support the notion that the politics of infrastructure finance has changed substantively. As Figure 6.9 (real volume of nonschool bond proposals by year) demonstrates, bond proposals generally rose through the 1950s into the 1960s, with a sharp peak in 1968. They then began to drop in the 1970s, reaching a low point in 1980. The 1980s have seen some rebound, and a spike in 1988. Yet the overall pattern of recent years has remained well behind the average of the first two postwar decades.

The drop in the volume of bond proposals during the 1970s corresponded with a decline in public support. Figure 6.10 shows that bond

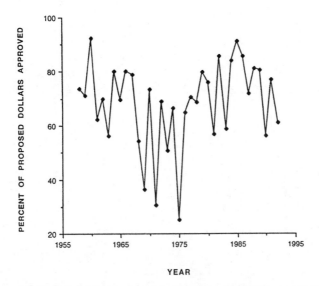

Figure 6.10. Voter Approval Rate for Nonschool Bond Proposals by Year
SOURCE: Data from the *Bond Buyer Municipal Statbook* (1986) and unpublished Bond Buyer statistics.

voting, "a tangible expression of public fiscal sentiment," indicated a concern over taxing and spending, with approval rates notably down from 1970 through 1975 (Anton, 1989, p. 151). But more recently the public has been willing to pass bond proposals and support infrastructure investment, particularly for transportation.

Together these trends suggest that the public is ready and willing to invest in infrastructure. The challenge for public officials is to take the leadership in proposing to raise investment.

Again, New York State offers a case study of the trends. Bond issues are particularly important for New York, because the state's gasoline tax has been included in the general fund budget since 1968. Thus the state has long paid for highway capital investment with voter-approved bond issues, rather than simply through a dedicated tax or revenue stream. Figure 6.11 shows that the postwar period saw regular new bond issues (in constant dollars) every year through the 1950s and 1960s, with peaks of $750 million in 1968, 1971, and 1972. In all of 1974-1982, however, the state issued only $150 million in new debt.

What happened? New York relied on bond issues to pay for transportation capital investment, and that in turn required initiative on the

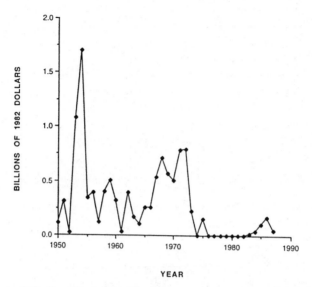

Figure 6.11. New York State Bond Proceeds for Highways by Year
SOURCE: Data from the U.S. Federal Highway Administration (1987, and subsequent issues).

part of its leaders and support from its electorate. In 1967 New York's voters faced a $2.5-billion transportation proposal and passed it. Four years later, another proposal by Governor Nelson Rockefeller, for a $2.5-billion bond program, went down to defeat. A $3.5-billion highway and mass transportation package, on the ballot in 1973, was also defeated. In 1977 Governor Hugh Carey failed to win approval of a $750-million proposal for economic development and job creation, as well as investment in basic facilities. The drop in new bond issues during the 1970s thus resulted from the public's antipathy to new spending and investment plans, some of which were intended for purposes broader than highways or mass transit.

More recently, though, the state's voters have generally supported infrastructure renewal. In 1979 the Empire State passed a limited proposal for about $500 million in capital spending. Four years later, Governor Mario Cuomo succeeded with a $1.25-billion bond proposal for rebuilding infrastructure. In addition, a 1988 investment plan for $3 billion received strong voter approval. Although infrastructure renewal has been well received by New York's electorate, debt for ill-defined economic development purposes has fared less well. Governor Cuomo's proposal for $800 million in job-creating public investment, on the state's ballot in 1992, went down to defeat, amid public concern over the details of spending and project choice. The real problem for New York, from 1973 to 1979 (and through the 1980s), was not so much public resistance to new investment as the lack of opportunity to vote on bond proposals targeted for infrastructure renewal. Moreover, although the state has resumed issuing debt to pay for infrastructure, the inflation-adjusted volumes of debt do not match those of earlier decades. Current needs, however, exceed those of past years. And proposals to raise the state's well below average gasoline tax to levels in line with neighboring states have regularly failed in the face of political divisions between Upstate and New York City, and between Democrats and Republicans.

Fiscal Innovation

The avoidance of tax increases and voter-approved debt has not crippled local spending for projects of high political priority, though. Indeed, the local fiscal environment is best characterized as one of "multiple cracks," with a variety of mechanisms for managing public development needs. Special districts, notably increasing in number, provide one means of dealing with infrastructure provision. Special road and highway districts, like those in Florida and Pennsylvania, offer a fiscal alternative. Even in post-Proposition 13 California, the voters have accepted a number of dedicated financing arrangements for infrastructure. For example, in 1984 Santa Clara County voters adopted a dedicated sales tax for highway construction. In addition, a new half-cent local sales tax, approved in 1988, is supporting new road and highway improvements in Ventura County, with more than $700 million anticipated over 20 years; San Jose now uses dedicated real estate taxes to support street and traffic improvements. Developers have taken the lead in dedicating rights-of-way and even the funding for traffic improvements across the nation. In older cities, like Cincinnati and Kansas City, the public has supported changes in state law and special tax increases to ease the funding for basic infrastructure needs.

The financing of new public facilities for economic development, such as convention facilities, arenas, and stadiums, also shows great local resourcefulness. Philadelphia is developing a new convention center complex, at a cost of close to $510 million, through a public authority and with substantial state financial help. San Jose, California, has built a new downtown convention center, a sports arena for its National Hockey League Sharks, and even loans of art work for the city's museum, employing the tax-increment funds generated by its local redevelopment agency. And even fiscally strapped Detroit managed to issue $180 million in revenue bonds for the expansion of the Cobo Center convention hall. Although financing public facilities is difficult and complex, with the political incentives for public investment, it can be done.

Policy Implications

States and localities obviously face concerns over tax increases and stringent budget situations. Still, when the public perceives a need for infrastructure investment, it is willing to support it with higher taxes and greater debt. Even in places with the severest economic and fiscal problems, like St. Louis and Detroit, local officials have implemented expensive projects that represent a high priority. The challenge for federal policy is to encourage local officials to expand infrastructure investment and to seek local support. Such policies might be politically difficult in the short run, but ultimately they could serve both national and local needs more efficiently and effectively.

■ Conclusions

Although many contemporary arguments about infrastructure need and financing requirements stress enormous structural and condition problems demanding a vast fiscal commitment, in reality the problems are far more modest and manageable. The rhetoric of collapse and underspending obscures the reality that problems are concentrated in a very few states or cities (Rohatyn, 1990, p. 8). Simply put, everything is not crumbling or collapsing, although some states and localities do face substantial bills for infrastructure repair and renewal.

Figures on capital spending tell the same story—there is no serious mismatch between needs and public spending. State and local capital investment for basic infrastructure effectively matches the peak spending levels of the 1960s. And recent increases in state gasoline taxes and federal aid will continue to fuel increased highway capital spending.

Moreover, to argue that the general public is indifferent to infrastructure needs, or unwilling to spend to meet those needs, is inaccurate. The recent history of bond issues on the ballot suggests instead a public quite willing to tax itself to invest in infrastructure. Annual bond approval rates (in dollar terms) averaged 77% from 1980 to 1989. Even in 1992 bond approvals came to 61%, with approval rates in infrastructure-needy states, like Missouri and Colorado, above the national average. And dedicated taxes for infrastructure have also received striking public support, as the recent passage of a gas tax increase in California attests. What is often missing is the political leadership and initiative required to offer the voting public the opportunity to support infrastructure investment, even at the cost of higher taxes.

At all levels of government, the difficulty is targeting resources to real needs. The "heavy artillery" of a trillion-dollar fiscal commitment and massive new federal aid makes little sense when needs vary so much from place to place. What does make sense is a "rifle shot" approach to infrastructure policy that matches specific policy and revenue changes to particular functional and area needs.

In the case of bridges, for example, it makes sense to encourage state and local governments to invest in the maintenance and renewal of existing structures, and to design a federal policy that rewards such state efforts. In contrast, current bridge renewal policies at the federal level provide such abundant capital assistance (80% of the cost of rehabilitation or replacement) that state governments face a disincentive to invest in maintenance with their own dollars.

A parallel policy mismatch occurs in federal highway aid. Though most of the rhetoric of infrastructure need has stressed physical deterioration and decay, state and local capital spending outcomes favor new construction and added traffic-carrying capacity. National policies that reward states and localities for their own investment in infrastructure renewal and maintenance would promote more effective public stewardship, while avoiding the dilemma of substitution.

A very different situation exists in the case of sewers and wastewater treatment facilities. There, user fees offer the best means of generating needed revenues and apportioning the true costs of service. Of course, some places, small towns for example, may lack the resources to manage clean water goals in the short term. For them, a combination of grant support and technical assistance may be necessary from both federal and state governments.

Finally, some of the most visible infrastructure needs and problems occur in older urban centers. Yet at the same time some city governments have managed to spend millions on convention centers, downtown improvements, and economic development. Federal urban policy should be structured to aid these cities in meeting their real needs, while promoting greater local attention and investment in basic infrastructure. Perhaps a return to federal categorical grants, specifically aimed at infrastructure renewal but with a requirement for local financial matching, could promote greater responsibility on the part of city governments.

Our current infrastructure needs are by no means insurmountable. They do require that we make some choices about how much we raise in taxes and revenues and what our spending objectives are, and that we recognize the complexity and variety of the intergovernmental system that manages and supports our infrastructure.

NOTES

1. For more on this point, see U.S. Congress, Congressional Budget Office (1988, pp. 130-140).

2. The relationship between public infrastructure and national economic output has received considerable attention in recent years. One analysis stresses the link between the volume of public capital stock and multifactor productivity. This statistical approach, exemplified by the work of economist David A. Aschauer, concludes that the declining stock of public capital explains the productivity slump of the 1970s and early 1980s, and thus offers a means of improving our economic "competitiveness" through substantial new investment in infrastructure (see Aschauer, 1988).

An alternative view holds that although Aschauer's analysis demonstrates a strong statistical relationship over the post-World War II period, it provides a weak guide to future policy, with little substantive meaning. It should be noted that the infrastructure-productivity relationship appears less strong during the 1980s, and depends upon estimates of public capital stock that are currently being revised upward. Indeed, "it is hard to imagine where one would find room for or what one would buy with the nearly $5 trillion of core infrastructure that Aschauer's estimates suggest we need to bring public investment into balance with private capital" (Aaron, 1990).

3. Some infrastructure analyses have noted that although real investment has now returned to the levels of the late 1960s, the national economy is now much larger. Thus, there is currently less capital investment relative to GNP than in the past, a point that has been used to support the contention that spending must be boosted dramatically. The assumption in this argument is that the *need* for public capital investment is a direct function of the size of the economy. But as the composition and functioning of the economy change, so does the need for infrastructure, both in total and in individual functions. The increasing shift to a service economy has probably reduced the volume of

public capital needed to sustain economic output. The fax machine, for example, reduces the need for the physical movement of documents, and thus the demand for courier and transportation services that claim highway and airport capacity. In turn, there is increasing demand for private investment in telecommunications capital.

4. This discussion of capital spending trends employs data on *state and local government capital spending, including federal aid.* These governments are responsible for almost all capital spending for basic infrastructure, and for implementing federal grant programs in such areas as highways and wastewater treatment facilities.

5. Given the varying character of highway systems and the more limited traffic-carrying role of off-system bridges, these cross-state comparisons focus on federal-aid system bridges.

6. Federal Highway Administration figures classify engineering and right-of-way costs separately from new construction, reconstruction, widening, and bridge renewal. If an appropriate proportion of these engineering and land costs were allocated to new construction, the actual percentages would be higher.

7. Data on convention centers from a survey by Gladstone Associates, Washington, DC.

8. Unpublished data provided by the Construction Division, Bureau of the Census, U.S. Department of Commerce.

9. Data on new bond issues from a special tabulation by IDD Information Services, New York.

REFERENCES

Aaron, H. (1990, June 28-29). *Comments on "Why is infrastructure important?" by David A. Aschauer.* Paper presented at the Federal Reserve Bank of Boston conference, Harwichport, MA.

Anton, T. (1989). *American federalism and public policy.* New York: Random House.

Aschauer, D. (1988, September 13). R_x for productivity: Build infrastructure. *Chicago Fed Letter, 13.*

Bond Buyer. (1986). *Bond Buyer municipal statbook.* New York: Author.

Choate, P., & Walter, S. (1981). *America in ruins.* Washington, DC: Council of State Planning Agencies.

Gramlich, E. (1977). Intergovernmental grants: A review of the literature. In W. Oates (Ed.), *The political economy of fiscal federalism* (pp. 219-239). Lexington, MA: Lexington Books.

Hornbeck, J. F. (1990). *Maintaining highway and bridge investments: What role for federal grant programs.* Washington, DC: Library of Congress, Congressional Research Service.

Meyers, H. G. (1985, July). *Displacement effects of federal grants to states for the primary, secondary and urban federal aid highway systems.* Washington, DC: Special Studies Division, Office of Management and Budget, Executive Office of the President.

National Council on Public Works Improvement (NCPWI). (1988). *Fragile foundations: A report on America's public works.* Washington, DC: Author.

Rohatyn, F. (1990, April 12). Becoming what they think we are. *New York Review of Books, 37*(6), p. 8.

U.S. Commission to Promote Investment in America's Infrastructure. (1993). *Financing the future*. Washington, DC: Government Printing Office.

U.S. Congress, Congressional Budget Office. (1988). *New directions for the nation's public works*. Washington, DC: Government Printing Office.

U.S. Department of Commerce. (1978). *Local public works program—Status report*. Washington, DC: Government Printing Office.

U.S. Department of Commerce. (1990). *Local public works program—Status report*. Washington, DC: Government Printing Office.

U.S. Department of Commerce. (1993). *Local public works program—Status report*. Washington, DC: Government Printing Office.

U.S. Department of Housing and Urban Development. (1992). *Report to Congress on community development programs*. Washington, DC: Government Printing Office.

U.S. Department of Transportation. (1990). *The status of the nation's highways, bridges, and transit: Conditions and performance*. Washington, DC: Government Printing Office.

U.S. Department of Transportation. (1992). *The status of the nation's highways, bridges, and transit: Conditions and performance*. Washington, DC: Government Printing Office.

U.S. Department of Transportation. (1993). *The status of the nation's highways, bridges, and transit: Conditions and performance*. Washington, DC: Government Printing Office.

U.S. Department of the Treasury. (1985). *Federal-state-local fiscal relations*. Washington, DC: Government Printing Office.

U.S. Federal Highway Adminstration. (1987). *Highway statistics summary to 1985*. Washington, DC: Government Printing Office.

U.S. General Accounting Office. (1988, May). *Bridge condition assessment: Inaccurate data may cause inequities in the apportionment of federal-aid funds* (GAO/RCED-88-75). Washington, DC: Government Printing Office.

Voters gave nod to $11.7 billion of bonds in '89. (1990, January 19). *Bond Buyer*, p. 1.

◾ Appendix: Trends in Broad Definition Infrastructure

The analysis in this chapter has dealt with basic infrastructure, including highways and streets, sewers and wastewater disposal, and water systems. Other discussions of national infrastructure spending and need have employed a broader definition of public capital facilities, including mass transit, aviation and airports, railroads and water transportation. The Congressional Budget Office's assembly of an expenditure database on broad definition infrastructure provides a basis for comparing capital spending trends between these two general categories of infrastructure investment.

The overall trend of broad definition infrastructure spending from 1956 is shown in Figure A.1. In real dollar terms, total spending rose

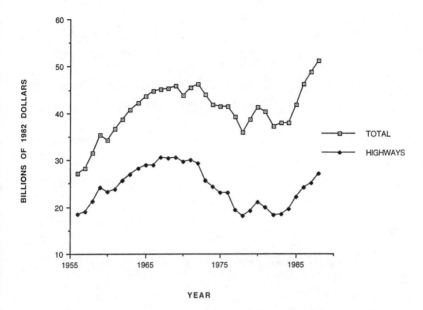

Figure A.1. Broad Definition Infrastructure Spending in Constant Dollars by Year

steadily through the 1950s and the 1960s, reaching a peak in 1972 of slightly more than $46 billion. There was then a clear decline to about $36 billion in 1978. The past few years have shown a striking increase. Spending grew from slightly less than $38 billion in 1984 to more than $51 billion in 1988.

The postwar trend of broad definition infrastructure thus tells the same story as the more limited category. *Current spending has sharply rebounded from the drop of the 1970s, and today is comparable to the peak years of the 1960s.* This pattern does not indicate either a lack of investment in infrastructure in the aggregate or pervasive public inattention.

The trend of infrastructure investment is also dominated by a single activity—spending for streets and highways. Highways now account for more than half of broad definition infrastructure investment, and the spending trend has largely been defined by highway investment levels. Again, the decline of the 1970s has been followed by a sharp upward trend in the past few years.

7

Building the City Through the Back Door: The Politics of Debt, Law, and Public Infrastructure

DAVID C. PERRY

The impasse is deep, Americans' appetite for government services exceeds their willingness to be taxed.

Robert J. Samuelson, *Washington Post*

Building the public city in the United States is part of everyday life and politics. The process includes services as proximate as today's potholes and as futuristic as tomorrow's information superhighway. It also triggers that most fundamental of questions in American politics: "Who's going to pay?" No one likes potholes, and almost everyone is intrigued by the possibilities of a cable-ready future; but such convergence of concern and interest breaks down when it comes time to pay the bill. The growing number of potholes in the street is just as likely to be the result of lack of funds for maintenance as the result of inclement weather. And although there is a growing consensus that every household can benefit from the services generated by the fiber-optic cable and broadband linkages, the debate over ownership and finance (among cable companies, telephone companies, and the state) actually outstrips the concerns over universal access and citizen rights.

These contemporary demands for service and quarrels over funding are not new. Demands for new public works—whether they be based on the economic conditions of trade that stimulated the canals, or the spread of typhoid and diphtheria that brought on the need for new sewers and water treatment systems—have always induced the systemic integration of new technologies into the social and economic transformation of the city. In the past, the magnitude of these new demands and the complexity of their requisite new "technological sinews" or public works networks (Tarr & Dupuy, 1988) combined to generate costs that challenged and, more often than not, temporarily overwhelmed the resources of the private and public sectors (Anderson, 1988; Seeley, 1993; Taylor, 1951). Building the public city has always been as much an issue of finance as it has been an issue of technology transfer in the service of a social or economic need.

The issue of funding is not simply one of amount, it is also about *who* should pay. In a liberal-capitalist state, should not infrastructure be first and foremost a product of the market? If not, what are the constitutional and political limitations on the intervention of the state? The history of the search for solutions to these questions is the day-to-day practice of what passes for public infrastructure policy in the United States. This chapter is an exploration of some key examples of this practice; uncovering, along the way, a fragmented and fitful mixture of public-private, as well as intergovernmental, relations (Goodrich, 1960).

■ The Federal Government: A Study in Reluctance

From the early days of the republic, the federal government has provided an inconsistent approach to the provision of public infrastructure. George Washington and Alexander Hamilton were among the first federal leaders to give strong support to "internal improvements," such as turnpikes and canals. Although their support of public infrastructure projects was followed by a full-scale plan for canals and turnpike networks in 1806 by U.S. Treasury Secretary Albert Gallatin, the first half of the nineteenth century was a period of substantial debate over the role of the federal government in support of public works. The debate was fueled by three distinct issues: (1) whether the federal government had the constitutional authority to build highways and canals; (2) whether the federal government could fairly intervene in the highly charged environment of regional and state competition to secure dominant positions in the mercantile networks of the growing country; and (3) what the proper role of the government was, in general, in the "taking" of land through eminent domain.

Proponents of direct federal participation in the construction of the nation's public works argued that such intervention was constitutionally consistent with the responsibility of the federal government to "promote the general welfare." Opponents contended that the building of internal improvements exceeded the role set forth for central government in the Constitution: Nowhere in the Constitution was it written that the federal government should build roads and canals. In fact, Andrew Jackson, elected on a populist and states' rights platform, viewed the internal improvements movement as a double-edged threat to democracy in that it sought to privilege the big-moneyed interests of business development and concentrate altogether too much power in the hands of political leadership in Washington. As a result Jackson vetoed $100 million of proposed federal involvement in canals, turnpikes, and railroads.

The Jacksonian position on the use of federal funds to build public works held, with a few notable exceptions, for the remainder of the century. However, for all the vigor of Jacksonian populism and the rigor of constitutional constructionists, federal involvement in public works most often fell victim to "regional jealousies" (Seeley, 1993, p. 22). During the intense economic competition over establishing the most efficient roads and canals and, later, railroads, federal intervention that privileged one regional highway over another was hotly contested. It was not simply, as Jackson argued, the case that the federal government should not serve the interests of big business in general, it was that it *could not* serve one set of business leaders without disadvantaging another region and business group. The ensuing struggle for such federal patronage muddied the entire federal role in public works. For example, when Congress, using parts of the Gallatin Plan, attempted to build a "National Road" in 1820, state rivalries, deal-making, and congressional wrangling produced a road that was only partially finished and still not fully funded almost 20 years later.

By the 1850s almost every major domestic issue to reach the floor of Congress was colored by the intense regional conflict between the North and the South. This included public infrastructure issues such as the federal role in determining the location of the eastern terminus of the transcontinental railroad. National politicians from the South, in particular, were opposed to the federal government's having a role in this or any other internal improvement matter, "fearing that such programs would strengthen the central government and create a precedent for federal intervention in other state matters, meaning, of course, slavery" (Seeley, 1993, p. 23).

These issues of regional politics, populism, and constitutional construction went a long way toward setting in motion the now longstanding tradition of resistance to an active and well-defined national public works policy. These issues derived strength from an even more fundamental early debate over the role of the liberal state in the determination of what was "public." The Supreme Court, as early as 1795, recognized the authority of government to "take" private property "when state necessity requires." The Court argued that this authority "exists in every government; the existence of such power is necessary; government could not exist without it."[1] This notion of eminent domain—the authority of the government to appropriate private property when it was deemed to be in the public interest—is at the heart of the American ambivalence over the provision of public works and the protection of individual freedom. From the very beginning of the legal debate over eminent domain, there were those who would use the principle to extend the states' role to take private property in order to create, as many developers in the new cities of the United States wanted, the "public" works to further development and meet the substantial needs of a diverse and growing population. On the other hand, there were those who feared that this practice might ultimately "render all private property subject to eminent domain takings by legislative whim. The accelerating pace at which states and municipalities were demanding internal improvements and public works presaged ever-increasing infringement upon private property rights" (Schultz, 1989, p. 40).

This latter concern was expressed clearly by Daniel Webster in the famous 1848 Supreme Court Case, *West River Bridge Co. v. Dix,* in which he argued that if the agents of government and, by expansion of their authority, legislators "are to be the sole judges of what is to be taken, and to what public use it is to be appropriated, the most leveling ultraisms . . . may be successfully advanced."[2] The Court rejected this concern of heightened egalitarianism by Webster and ruled that the government does have broad authority to take land for all "public" purposes.[3] As a result, though the federal government effectively withdrew for most of the century from an active fiscal and programmatic role in public infrastructure policy, it clarified the practices of eminent domain and, through these powers, "took" literally millions of acres [4] of land to support the development of the railroad, turnpike, and canal systems.

Therefore, the early role of the national government in building public infrastructure was advanced and restrained at the same time.

From the beginning, public works policy would be a decidedly mixed effort that dampened the active fiscal involvement of the federal government, but advanced, with eminent domain, the definition of "public" property and the purposive construction of internal improvements and public works on such property.

The States' Role

But if the nineteenth century was to be the time when participation of the federal government in the provision of public works became increasingly limited, it was also the time when the *active* participation of states and localities expanded dramatically. Charles Francis Adams,[5] in an 1840 essay on the rising amount of state debt used to build public works, suggested that this change in the level of governmental assignment had substantially altered the practice of public affairs in America. In what may be considered one of the first descriptions of the institutional framework of America's "backdoor" practice of building public infrastructure, Adams (1892, p. 324) wrote:

> The Americans are proverbial for never being discouraged. If they can not carry a point directly, they will manage to do it by some roundabout way. They were determined upon improving communications between the seaboards and the interior. A very large number of them thought it best that this work should be done under the super-intendence of a common head, and they proposed a method of action accordingly. But the extent of it excited the apprehensions of a still greater number and they refused to adopt it. According to them, the business could be intrusted with safety only to the care of the separate States upon whom it was in the end devolved.

Adams was writing at a time when states were entering into the construction of public improvements with unbridled enthusiasm (New York State Constitutional Convention Committee [NYS CCC], 1938b, p. 85). In 1835 French transportation specialist Michel Chevalier, writing from Buffalo, New York, after viewing the Erie Canal, remarked on the intensity of road and canal construction going on in the United States:

> The spectacle of a young people, executing in the short space of fifteen years, a series of works, which the most powerful States of Europe with a population three or four times as great, would have shrunk from under-

taking, is in truth a noble sight. The advantages which result from these enterprises to the public prosperity are incalculable. (Chevalier, 1839)

Chevalier was referring to what he perceived to be the doubly difficult conditions of building public works in America, as compared to Europe. First, there was the sheer size and undeveloped character of the country—the distances between places were far greater than almost any place in Europe, the mountains to be crossed were higher, the terrain wilder and the rivers and lakes larger. All of this combined to make canal construction a problematic process and a costly one. Europeans were used to smaller projects in far more understandable environments and at costs that, from past experience, could be predicted. As a result, at a time when internal improvements projects in Europe were often contained enough to be profitable private enterprises, the staggering magnitude of the tasks and the accompanying risks of frontier America precluded active and exclusive participation by the private sector. Besides, this was not a country that had yet produced a substantially wealthy investor class. As New York Senator William Seward observed: "[A] great and extensive country like this has need of roads and canals earlier than there is an accumulation of private capital within the state to construct them" (as quoted in Seeley, 1993, p. 23).

Therefore, building infrastructure in the United States, for all the constitutional and regional debate, had become a decidedly *public* enterprise. The combination of federal governmental reticence and limited private capital meant that building public infrastructure would be, as Charles Adams had observed, largely the role of localities and especially the states. In the highly competitive mercantile environment of frontier America, states really had no choice but to become actively involved in building and maintaining urban public works, as well as large-scale road and canal construction. With the federal government reluctant to raise funds either through taxes or by borrowing for public infrastructure,[6] the states willingly engaged in highly speculative ventures meant to support the construction of improved transportation networks: The number of miles of canals jumped from 1,270 in 1830 to 3,700 in 1850 and 4,254 in 1860, and the number of miles of railroads went from 70 miles in 1830 to 9,000 miles in 1850 (Isard, 1942; Segal, 1961, p. 172).

These ventures were speculative for two reasons: first, as Guy Callender (1903) suggested, there was a substantial paucity of private start-up capital in the new nation; and second, for all the economic promise represented in the new canals, state legislatures were loath to raise taxes to support them. As a result, new canals and later railroads and water systems were funded from the start through large-scale public borrowing. State debts rose quickly, from more than $12 million in 1820 to more than $200 million by 1842 (Adams, 1892; Quirk & Wein, 1971). Even more important than the size of public debt was the fact that most of it was unsecured. For example, in the wake of the dramatic success of its investment in the Erie Canal, New York State had amassed $38 million in unsecured debt by 1846. States began to build canals, turnpikes, and railroads all along the coast, with money they didn't have. "Many of these public works," concluded the New York State Constitutional Convention Committee (1938a, p. 84) "were the result of unrestrained enthusiasm rather than sound judgement." By 1861 more than $188 million had been invested in canals, three quarters of which was provided by governmental funding. Fully 90% of these governmental monies, or $127 million, was raised by states as they borrowed "through the sale of bonds in the national and international money markets" (Segal, 1961, pp. 180-181).

The construction of railroads, with their esoteric technologies and their dramatic new product costs, created even more problematic financial practices. In the highly competitive and entrepreneurial days of the first railroads, states actually gave their credit away to private railroad companies, under the belief that support for private rail enterprise was necessary because it appeared not to be competitive with canal transit. New York State led the way in this regard, issuing $3 million of stock to the New York and Erie Railroad to help its construction. The state went further and backed up the stock with the "full faith and credit of the people of the state" (Quirk & Wein, 1971, p. 527), feeling secure in the fact that the railroad would be able to service the debt out of revenues. When the economy crashed and the Panic of 1837 set in, the railroad defaulted on its payments and the state attempted to reclaim the assets of the railroad, only to find itself able to reclaim just the thin strip of property the track was set in and the track itself; the rolling stock, buildings, and debts were all still the property of the private company.

A string of similar failed investments forced New York and other states "to the very brink of dishonor and bankruptcy."[7] The people of Michigan, for example, were led by an equally political and entrepreneurial legislature. They had kept taxation in the new state down to $.70 per head, and yet the drive for internal improvement had enticed them through the door of constitutionally uncontrolled debt practices, so that the cost of servicing just the interest on their loans had reached $1.25 per head, and the revenue streams of many of their key internal improvements were threatened or erased in the depressed, mid-nineteenth-century national economy (Adams, 1892, p. 332). In all the states, the amount of unsecured debt for canals, railroads, banks, and roads had risen to more than $207 million by 1842. Governor William Bouck of New York spoke for most state leaders when he lamented, in his 1843 annual message, that

> The Legislature of 1842 convened at a period of great embarrassment in the financial affairs of the State. The treasure empty, our credit seriously impaired; the State stocks were selling at ruinous sacrifices; temporary loans were nearly at maturity; the time for the quarterly payments of interest on the public debt, amounting to more than $2,000,000 was fast approaching; contractors were pressing for payment, and the progress of public works virtually suspended.[8]

In one legislature after another, states started to retrench their fiscal approach to public infrastructure. In fact, no single issue did more to trigger constitutional reform in states than that of the abuse of the states' powers to create indebtedness (Adams, 1892; NYS CCC, 1938a; Quirk & Wein, 1971). More particularly, Henry Adams argued it was the "failure of this policy of internal improvements . . . [that] led to radical changes in the constitutions of the various States" (Adams, 1892, p. 340).

In no state was this more clearly the case than in New York. With the rising debacle of unpaid debt facing the legislature, and with a determination on the part of the legislators to pay off this debt, even if it meant raising taxes to do so, the members of the legislature themselves set about to change the rules regulating borrowing. In a bill that became known as the "People's Resolution," legislative representative Arphasad Loomis sought to bind the legislature to an agreement to create debt only for projects that had first been approved in a referendum by the voters of the state (NYS CCC, 1938b). His bill was defeated in a tie vote in the Assembly, but a version of debt by binding referendum went on to be a part of the constitution produced by the New York State Constitutional Convention of 1846. The constitution was ultimately changed so that a referendum was required for the creation of any debt of more than $1,000,000 (Art VII, Sec 12). Proponents of this restriction argued: "We will not trust the legislature with the power of creating indefinite mortgages on the people's property. . . . And . . . that whenever the people were to have their property mortgaged for a State debt, that it should be done by their own voice, and by their own consent" (as quoted in Quirk & Wein, 1977, p. 533).

By 1860, 19 states had adopted constitutional amendments limiting debt, and the constitution of each new state entering the republic contained some form of debt limit provision. As a result, by the turn of the century, state debt had actually declined. And today there are a variety of arrangements that can be found in various states. In 19 states the state is required to submit general proposals or obligations to a popular referendum. Twenty-one other states impose a limit, either through a debt ceiling or through a percentage of property tax values, on the maximum a state is allowed to go into debt (Walsh, 1978, pp. 19-22). Nine states impose no constitutional limitations and rely on the state legislatures to control state borrowing, although many of these states tend to have had, in the past, relatively low debt levels (Walsh, 1978, p. 23).

■ Building Public Infrastructure Through the Local Back Door

Constitutional restrictions, such as the 1846 binding referendum clause of the New York Constitution, may have had conservative consequences on state borrowing (Lincoln, 1900, Vol. II), but they soon stimulated a rash of activities at the local level. In the process, state legislatures took a page from the book written a generation earlier by the national government. Although they tightened up the constitutional regulations on debt formation at the state level, they loosened their regulations on debt formation at the local level. In almost all states the seeming inconsistency or "fancy" (Adams, 1892, p. 357) of this practice—where states could constitutionally restrict themselves from certain debt practices in the issuance of bonds for railroad construction, and at the same time allow local governments, themselves creatures of the state, the license to issue such bonds—was challenged in the courts. Except for Michigan, most states, from Iowa to New York, supported the politics of this practice. It was "not, then, an accident that the expansion of local credit took place almost immediately after States had been shoved off the stage of industrial action" (Adams, 1892, p. 357). So although states practiced a legal policy of fiscal restraint and limited taxation, they instituted a politics of fiscal largesse and flexibility at the level of the city and other local jurisdictions.

In short, cities were now empowered to do what state legislatures had constitutionally prohibited themselves from doing: create the financing with which to build public works without either raising taxes or asking for the approval of voters. Just as states had been the "roundabout," as Charles Francis Adams had called them, or the "back door" through which national internal provisions had been built in the early nineteenth century, by late in the century cities were becoming the governmental door through which state and local political leaders would slip their fiscal strategies for building the public city.

This reversal of roles led to new fiscal patterns that were as dramatic as they were sustained. In 1840 the total debt of all local jurisdictions in the country was only $25 million, but by 1880, the bonded debt of all localities was more than $833 million (Adams, 1892, p. 305); and by the turn of the century, local debt had almost doubled again, totaling more than $1.5 billion (Walsh, 1978, p. 19). The debt incurred by the larger cities in 1880 comprised 82% of the total local debt, or $682 million. Total state bonded indebtedness, which had been just a bit more than $207 million in 1842, only climbed to $246 million in 1880, and actually declined to $235 million in 1900 (Adams, 1892; Walsh, 1978). What is even more interesting is that, although almost all state debt in the 1840s was attributable to public infrastructure, the amount of debt incurred by states to build railroads, canals, and other internal improvements had dropped to $56.5 million by 1880—a direct result of constitutional provisions on debt formation for public improvements (see Table 7.1).

In the cities of the country, however, the exact opposite was the case. Where almost no local funds had been expended on internal improvements in the 1830s and early 1840s, of the $682 million of big city debt accumulated by 1880, more than two thirds was committed to building not only railroads and canals, but also waterworks, parks, sewage systems, public buildings, schools, and libraries—the full range of technological and physical systems of the new industrial city.

It would be wrong to conclude that this shift in the location of public works policy to cities was simply the result of state legislative politics. The cities of America were growing faster than any other urban centers in the world. Wave upon wave of new immigrants from the countryside and from Europe combined with the industrial restructuring of the urban economy to generate demands for public infrastructure that were un-

TABLE 7.1 Amount of Bonded Debt in 1880, for States, Cities and Minor Civil Divisions, and Purposes for Which Bonds Were Issued

Purpose for Which Debts Were Contracted	State and Local Indebtedness as Given in Census of 1880	Indebtedness of Cities With Population of 7,500 and Upward, as Given in the Census of 1880	Indebtedness of States as Computed From Balance Sheets of States	Indebtedness of Towns and Minor Civil Divisions Computed From the Foregoing
Bridges	$24,853,388	$20,809,431	....	$4,043,957
Cemeteries	283,816	272,912	....	10,904
Fire department	2,514,082	2,214,924	....	299,158
Public parks	40,612,536	40,490,636	....	121,900
Sewerage	21,370,536	21,335,434	....	35,102
Streets	86,674,860	81,502,817	....	5,172,043
Waterworks	146,423,565	141,797,828	....	4,625,737
Bounties, militia, war	75,154,400	28,722,787	$33,310,738	13,120,875
Funding of floating debt	153,949,095	122,864,804	2,978,048	28,106,243
Refunding old debt	138,743,730	71,071,140	57,057,862	10,614,728
Public buildings	48,493,952	25,516,829	6,327,780	69,345,365
Railroads	185,638,948	68,309,493	47,984,090	69,345,365
Canals, rivers, water power	36,224,548	16,726,064	8,655,780	10,842,704
Schools and libraries	26,509,457	13,889,915	....	12,619,542
Miscellaneous	130,374,758	26,571,446	90,000,000	13,803,312
Total	$1,117,821,671	$682,096,460	$246,314,298	$189,410,913

SOURCE: Adams (1892).

precedented in the experience of other nations, much less in that of the United States. Between 1790 and 1880, the number of cities with more than 7,500 inhabitants grew from 6 to 286, and the number of residents in these cities increased from 131,472 to 11,318,547. Overall the population of all people living in cities in 1880 was 14,129,735, and by 1920, this population had increased by almost four times this number, to 54,253,280. In short, by 1920 America had become a "nation of cities" (Glaab & Brown, 1967)—with more people living in municipalities than anywhere else.

Historian Stanley Schultz suggests that the waves of immigrants were greeted by other waves of epidemics, particularly, cholera and typhoid fever:

Over the first six decades of the century, urban death rates from all diseases apparently rose dramatically. Between 1815 and 1839 the average annual death rate of all citizens in Boston, New York, Philadelphia and New Orleans combined was 28.1 per thousand of population; from 1840 to 1860 the rate climbed to 30.2 per thousand. If one looked at the cities individually, the figures were even more startling. Cholera claimed nearly 150 victims per thousand in 1847; together the two diseases took more than 80 per thousand in 1853. In 1849 more than 55 people per thousand died of cholera in Chicago; five years later typhoid fever and cholera killed more than 60 per thousand. . . . Similar reports involving pneumonia, smallpox, tuberculosis, scarlet fever, and intestinal disorders came from other cities over the period. (Schultz, 1989, pp. 114-115)

The spread of these diseases was the direct result of the unhealthy condition of overcrowded, poorly maintained, and refuse-filled streets, coupled with massive increases in untreated garbage, wastewater, and human waste.

Meeting the challenges of such conditions required not only a fundamental rethinking of the fiscal relations of public infrastructure policy at the state level but also a whole new approach to policy at the local or urban level. Public infrastructure policy was not simply a trigger of growth, as it had been in the earlier decades of the century, it was also recognized as a response to the ravages of urban growth:

An increasing number of Americans had . . . come to believe that the solution to environmental problems lay in physical and technological innovations. . . . They articulated a new concept of "public works" thereby giving physical expression to the changing social vocabulary of the century's emerging urban culture. That conception urged that the public itself, through its tax dollars and its political support for public works projects, take charge of its own destiny in building a safer, saner, and more sanitary urban environment. While publicizing such projects . . . new professionals—particularly landscape architects and engineers—became aware that the implementation and administration of fresh technologies would require alteration in the reach and scope of municipal government. (Schultz, 1989, p. 153)

Local government, however, was more than simply the new place for the fiscal pragmatics for public works delivery. The new sewer and water systems, highways and sidewalks, paving and parks and other new technologies of industrialism promised to rectify the health and safety abuses of economic growth and stimulate changes in the structure and politics of urban governance (Felbinger, this volume; Shultz, 1989). The faster cities grew, the larger and more complex the systems of public infrastructure became—all of them requiring enormous amounts of new capital. This of course meant even greater increases in municipal debt—for example, the per capita amount of municipal debt doubled in the 1860s alone. And by the 1870s, during which time the nation suffered its second economic depression of the century, municipalities began to default on their loans, and states were forced to apply ceilings on municipal debt formation in the wake of serious financial failures in the local public sector.

From General Obligation Debt to Revenue Bonds

By the turn of the century constitutional limits on the general obligation debt of municipalities existed in all but a handful of states. Defenders of these limitations argued that they would prevent unsound investments and abuse of municipal borrowing powers due to "extravagance and corruption: while still enabling the municipality to undertake its governmental functions" (NYS CCC, 1938a, p. 295). Indeed, between the 1890s and the Great Depression of the 1930s, the municipalities were able to meet the exigencies of demographic growth and economic development with fairly significant amounts of new infrastructure in spite of the limits of the constitutional amendments. In most states, the dramatic changes in the urban climate were more than enough to stimulate a pragmatic legislative politics that produced constitutional regulations that were incredibly "complex and riddled with exceptions and inconsistencies. They represent[ed] a series of ad hoc responses by legislators to diverse problems" (Walsh, 1978, p. 23). For example, though the state put a debt ceiling on the expenditure of funds by general purpose governments in New York, the legislature created exemptions from debt limits for municipal water debts and (later) for the particular debts incurred for the construction of New York City's subway system (NYS CCC, 1938a: pp. 297-303).

Just before the turn of the century, the states began to encourage a new form of debt finance to both get around and parallel general obligation debt-financing of public works. The increasing pressures of urbanization and industrialization forced states more directly back into the arena of public works finance. For almost three decades most states had adhered to an essentially "pay-as-you-go" approach to building public works. But, beginning with water systems, states like Washington, in 1897, authorized the financing of the city of Spokane's water system strictly from bonds secured against the future revenue generated by the system. Two years later the state of Illinois authorized the creation of debt through the issuance of bonds payable solely through user fees. "The courts," observed the Council of State Governments (CSG, 1970, p. 5), quickly "developed a doctrine that revenue bonds payable out of special funds were not state debts within the meaning of constitutional and statutory debt limits."

For the states and their localities the revenue bond became the newest tool in the roundabout or backdoor approach to building public enterprises, or revenue-generating public works. With the promise of revenue, taxes need not be raised. The promise of revenue also created a projected source of capital with which to erase future debt ceilings. Revenue bonds helped states and localities around the onerous restrictions of constitutions and the heated politics of bond referenda. Instead of going to the public for funds, these governments were able to go to private-sector bondholders. The costs of the public work, it was argued, need no longer be borne by the taxpayers, but by the users. And the letting of the bonds could be "more efficiently" tied to times when they were needed most and their costs were the lowest. Further, it was argued that these bonds could be more effectively amortized through their systematic attachment to revenues (CSG, 1970, p. 6).

Although revenue bonds became a substantial new tool in the fiscal kit of states and localities, the chief debt instrument remained the tax-backed general obligation bond (GO bond), until the 1960s. In fact, the revenue bond only represents half the twentieth-century fiscal public infrastructure process. The other half is the "public authority." In the early 1960s two thirds of all state and local debt was still guaranteed through constitutionally restricted, tax-supported GOs. But by 1990, of the $893.6 billion of outstanding long-term state and local debt, 67.3 percent, or $601.9 billion, was secured through new forms of special agencies or public authorities who utilized the revenue bond as their main debt instrument (Axelrod, 1992). The revenue that secures the bonds differs in various states—user fees, charges, subsidies, or special taxes. However, almost all of this debt "carries no binding constitutional guarantees of repayment" (Axelrod, 1992, p. 19). Although, in many states, there is a moral obligation of the state or locality to pay back the bondholders in the case of default, there is no guarantee that the bonds will be backed by the "full faith and credit" of the state or the locality (Axelrod, 1992, pp. 4-7; Quirk & Wein, 1971, p. 570).

These uneven and highly politicized practices of limiting municipal general obligation debt and encouraging revenue bond financing first became evident in the 1930s, with the advent of the Great Depression. Although the economic conditions of debt crisis and the conflictive conditions of intergovernmental relations surrounding the financing of public works were similar to circumstances of past eras, the approach of Depression Era leaders to finding a way of addressing public infrastructure and socioeconomic change was different. During the nineteenth century, a depression led to calls to restrict state or local debt financing practices to ensure a measure of fiscal stability and rationality. As a result, governments simply shifted the responsibility and constitutional fiscal flexibility from one *level* of government to another. Constitutional reforms were thereby enacted in an attempt to limit dangerous, corrupt, or irresponsibly speculative government debt practices at one level, while ignoring similar threats at a lower level of government. In the 1930s attention was focused on finding ways to *overtly* circumvent the constitution(s) in order to *enable* local governments to continue debt financing of public works during a period of severe fiscal and economic crisis.[9] There was no level of government, previously still removed from the fiscal practices of public infrastructure, to turn to. All levels of government were equally threatened by the frailty of the economy in general and the overly subscribed conditions of debt in the public works sector at the local level. Therefore, what governmental leaders at all levels did, in the absence of another level of government to which they could assign the debt production function, was to turn to another *type* of government—the nascent public benefit corporation or public authority. Revenue bonds and this relatively new special purpose public corporation became the key ingredients in the next round of building America through the back door.

The Depression and the Rise of the Public Authority

In the first months of his administration, Roosevelt moved to attack the full range of economic conditions furthered by the Depression. More particularly, he viewed a dynamic domestic public works program as a multiple instrument of recovery: providing municipalities and states with new economically and socially stimulative infrastructure, and citizens and business with jobs and contracts, respectively. Although the economic payoffs of such public works projects were clear, the debt-ridden condition of most cities and states was such that their participation in these programs would have been highly problematic. Municipalities were fast approaching, if not already at, their debt ceilings. In fact, between 1929 and 1937 more than $3 billion in state and local debt, or 15% of the average municipal debt, could not be paid. More than 3,200 local governments were unable to make timely payments on the interest or principal of their existing debt (Perry, 1990a). Coupled with this was the long and storied constitutional history whereby states had spent nearly a century securing some modicum of fiscal control.

In order to accomplish his public works program, Roosevelt set in motion a clear and unambiguous set of procedures designed to circumvent "the normal [constitutional] restrictions that the states had come over the years to impose on borrowing by local governments" (Smith, 1964, pp. 108-109). Quite bluntly, Roosevelt had set about to overturn 100 years of history of the states' evolution of constitutionally prescribed ceilings on bonding debt and ad valorem tax percentages coupled with voter approval, by referendum, of dedicated bond issuances. He used the Reconstruction Finance Corporation (RFC) and Public Works Administration (PWA) for the express purpose of purchasing *revenue bonds* from a new, ostensibly ad hoc emergency agency of municipal government called the public authority. This agency, wrote Roosevelt (in a letter to the governors of the 48 states), would allow the municipalities to "legally . . . take full advantage" of what would ultimately be $1.5 billion of bonds. It was a new agency, that was

> essential, at least for the balance of the existing emergency . . . because it allowed the general purpose municipality additional powers . . . to undertake such projects and issue bonds to finance the same: thus escaping the difficulty of gearing the legal machinery which has served municipalities of your States adequately for decades to the speed with which the Federal Government must extend credit to achieve desired results. (Smith, 1964, p. 108)

The public authority, statutorily established in the legislative tradition of the federally established Port of New York Authority and the Tennessee Valley Authority, served as a financially welcomed and constitutionally acceptable back door through which to finance public works.

Hence, in the early 1930s the country launched a federal government-public authority nexus. States created these authorities in one of two ways: either by special legislative statute, as in the case of the Triborough Bridge Authority (TBA) in New York (Lines, Parker, & Perry, 1986, pp. 231-256), or by general act of the legislature where local authorities could be created by election decree or other means at the local level.

The statutory, special purpose authority became an immediate success. States, like Pennsylvania and New York, created literally scores of public corporations capable of selling revenue bonds to finance everything from highways to bridges and tunnels. By 1946 all but "seven states had adopted legislation for public enterprises to sell revenue bonds. Twenty-five of these states enabled local governments to create public authorities by ordinance" (Walsh, 1978, p. 28).

The Popularity of the Public Authority: From Moses to Rockefeller to Cuomo

What had started out as "an excusable, even a laudable, scheme in an emergency . . . with a purpose ordinarily limited to a single money-making enterprise" (Davis, 1935, p. 328), and geared to go out of existence once the project was built and its bonds retired, was quickly turned into a permanent part of the public sector. As the emergency conditions that allowed for the backdoor popularization of the authority faded, proponents argued that authorities should continue, not because they were debt-evading devices, but because they had emerged as highly efficient government agencies ruled by the "test of good corporate management" (Tobin, 1953, p. 16). Austin Tobin, the head of the powerful Port of New York Authority, argued that the complex urban environment necessitated a growing and vital public infrastructure built and managed by

> a public corporation set up outside the regular framework of federal, state and local government, and freed from the procedures and restrictions of routine government operations, in order that it may bring the best techniques of private management to the operation of a self-supporting or revenue producing public enterprise. (Tobin, p. 16)

There is not one shred of acknowledgement of the temporary, emergency-based roots of the authority to be found in this definition. Here was an argument for as pure a public reflection of the corporate model of the market function as a business or structural reformer could dream of (CSG, 1970, pp. 6-10; Schiesl, 1977). The statutory authority "clashed with the traditional concepts of fixed boundaries, real-property assessments and tax collections, elected representatives selected by the residents, and allocation of priorities of services" (Smith, 1964, p. 118) by federal, state or local government. Because of all this, the authority "recoiled within the separate function allotted it" (p. 118) and developed into the most self-sustained governmental unit the country had ever experienced.

In its 1953 study of public authorities, the Council of State Governments (CSG) defined them as a clear and unambiguous back door through which to fund costly or politically contentious public enterprises. In fact, it was their latent *extra*-constitutionality that the council celebrated in its definition of such authorities:

> Public Authorities generally are corporate bodies authorized by legislative action to function outside of the regular structure of state government in order to finance and construct and usually to operate revenue-producing public enterprises. Their organizational structures and powers are of the type usually associated with public corporations and like the latter they have relative administrative autonomy. Public authorities are authorized to issue their own revenue bonds, which ordinarily do not constitute debt limitations, because they are required to meet their obligations from their own resources. They lack the power to levy taxes, but are empowered to collect fees or other charges for use of their facilities, devoting the resulting revenue to payment of operational expenses and to interest and principal on their debts. (CSG, 1953, p. 113)

The popularization of the special purpose authority and its entrenchment as a permanent part of urban design and policy making was a direct breach of the letter and, in most cases, the spirit of the original use of statutory public authorities. The authority was born out of fiscal and economic expediency. Its circumvention of local and state debt limits, its political independence from traditional legislative oversight and public governance, and its direct operational control over public works projects were justified on the basis that such agencies were, as Davis (1935) suggests, temporary "stopgaps" legislated for an existence not to exceed the payback period of the bonds. When user fees and other revenues wiped out the debt represented by the bonds, the authority was to go out of existence and the public work was to become the infrastructure resource of the governmental body that had statutorily created the authority to begin with.

Perhaps the clearest example of the evolution of the public authority from a stopgap fiscal expediency into a permanent feature of American governance is the Triborough Bridge Authority. In 1925 the mounting use of the automobile and continued population growth required that New York City build another bridge across the Harlem River. By 1927 plans had been drawn up and on Friday, October 25, 1929, the day after "Black Thursday," building commenced. Three years and $5 million later, the bridge had yet to span the river, and the City had run out of money.

Robert Moses, in his capacity as chairman of the State Emergency Public Works Commission, viewed the public authority as just the tool to link economically troubled places like New York City to new federal sources of funding like the RFC and the PWA. In helping to write the original state legislation instituting the TBA, Moses and his aides stipulated that the TBA's "corporate existence shall continue for a period of five years and there after until all its liabilities have been met and its bonds paid in full." (New York Statutes, 1933, Chap. 145). Once the bridge was built, its bonds paid off from toll revenues and its ownership reverted to the City, the TBA was to go out of business.

According to the internal documents of the TBA,[10] Moses, who soon after the Authority was formed took over as director, strongly adhered to this formal definition of the public authority. He represented himself as being deeply worried about the revenue-generating capacity of the as-yet-uncompleted Triborough Bridge, and railed against the arguments of advisers who wanted to blend bonds to build the Bronx-Whitestone Bridge with the Triborough bonds.

During most of this early period, Moses remained financially very cautious. Only after long and heated internal debate was he willing to venture a constitutional change of the single-purpose nature of the public authority itself, transforming it into an agency that would build two bridges—the Triborough and the Bronx-Whitestone. Although he was willing to expand the Authority to build more than one project, he still kept the bonds from each bridge separate—so that the potential failure of one project would not jeopardize the fiscal soundness of the other.

After the Triborough Bridge opened, however, the notion of fiscal failure was quickly forgotten. Receipts from tolls exceeded all expectations, and it became readily apparent that the Authority would be able to retire the bridge's bonds at a faster rate than anticipated. Moses introduced another amendment to the State Constitution that altered the fiscal structure, just as the project structure had been altered in the previous amendment. But this was more than a simple refinancing agreement. It essentially guaranteed the existence of the TBA as a *permanent* agency of public works:

> When all liabilities incurred by the authority of every kind and character have been met and all its bonds have been paid in full . . . all rights and properties of the authority will pass to the City of New York. The authority shall retain full jurisdiction and control over all its projects, with the right and duty to charge tolls and collect revenues therefrom, for the benefit of the holders of any of its bonds or other liabilities even if not issued or incurred in connection with the project. (New York Statutes 1937, Chap. 3)

Moreover, the amendment goes on to stipulate that "the Authority shall have the power from time to time to refund any bonds by the issuance of new bonds, whether the new bonds to be refunded have or have not matured and may issue bonds partly to refund bonds then outstanding and part for any other corporate purposes" (New York Statutes, 1937, Chap. 3). This meant that any project consistent with the bridge-building mandate of the authority could be financed through a pooling of the bonds and revenues of the corporation.

Moses subsequently used the TBA and other authorities in a variety of ways—when one authority didn't work, he quickly abandoned it and designed a new one, or further redefined it to suit his needs. He was a genius at writing legislation (Caro, 1974) and in his use of his legal creations to produce a powerful new process for building public infrastructure. He proceeded carefully, never overstepping the fiscal potential of the authorities, working in one new step at a time. Indeed, at the height of his power he was the head of 14 different public works agencies and/or authorities.

These authorities were increasingly refined, becoming sophisticated, quasi-private corporations chartered by state law. Because their primary fiscal responsibility was to bondholders, they were not accountable to civil service dicta, ordinary line-agency fiscal responsibilities, and regular legislative oversight and accountability. Moses saw this arrangement instrumentally—not in terms of the logic of management. It was the fiscal freedom of the authority that was important to him.[11]

The public authority gave Moses three ingredients central to his success at institutional management: political independence, fiscal freedom, and full operational control over the entire project (Lines et al., 1986, pp. 231-256; Perry, 1990b; Spargo interview, 1983). In Moses's words, the public authority was "the warp on the loom" upon which the modern tapestry of twentieth-century New York was being woven (Moses, 1941).

Critics of the Moses approach to public infrastructure argued that the public authority, as Moses and others were employing it, had "degenerated into a debt-evading device."[12] This issue became a central feature of the Constitutional Convention of 1938. Moses defended the emerging approach of authorities, not as debt-evading devices, but as the opposite—"revenue-producing devices." "[T]he authority could be sustained by tolls. You can call that debt evasion if you please. I call it sound business."[13] He went further and argued that these devices, far from being unregulated public agencies, were regulated by the "most stringent" of agreements, made up of the covenants between the authority and the bondholders. For Moses, the authority was one way to "get things done": It generated revenue within the efficient and entrepreneurial boundaries of the market, and it did all this with a minimum of interference from the politicians or the people. Or, as Governor Al Smith put it, the authority is "the one method we have discovered of getting work done expeditiously and without overtaxing our people to get it done."[14]

In spite of the arguments of early critics and some modifications of the constitution, Robert Moses succeeded in using authorities to produce everything from parks, hydropower projects, bridges, highways, World's Fairs, housing and slum clearance projects, a baseball stadium, and the New York Coliseum. His success was the most dramatic example of the growing use of the authority model. From the end of World War II until the early 1960s, hundreds of similar public corporations sprung up around the country as the nation responded to the pent-up demand for public infrastructure, schools, and housing in the peacetime economy (Axelrod, 1992).

In the 1960s and 1970s the move of Americans to the suburbs and the Sunbelt transformed significantly the spatial and political demands for public works. The rise of special-purpose governments or districts to provide these infrastructure systems took on many of the characteristics of public authorities.[15] Literally thousands of these types of governments were created to accommodate the respatialization of urban America so that by the close of the 1980s, there were thousands of public authorities and between 29,000 and 35,000 special-purpose governments in the United States. These governments have taken on so many different forms and functions, in order to provide public service and evade the political pitfalls of raising taxes, that it is unclear how many actually do exist in the United States, and it is equally unclear how to arrive at a common definition of them all (Leigland, this volume).

What is clear is that states and localities, historically the key levels for the provision of public infrastructure (U.S. Congress, Office of Technology Assessment [OTA], 1990), have had to continue their innovative backdoor practices of building the public city. Again, New York State is not only a good example of these practices but also has led the way in pioneering the latest versions of this politics of the back door; namely, the "moral obligation bond" and lease-purchase agreements.

When he took office as governor of the State of New York in 1958, Nelson Rockefeller ideally should have been the most powerful man in the state—he had just been elected to the state's highest political office and, of course, he was a Rockefeller. But it was soon clear that he was not as powerful as he would have liked. Prior to his election the voters had blocked a housing bond issue for the third time, and right after his election voters turned down, for the second time, a highway bond issue and, a bit later, they turned down, for the fourth time, a higher education bond issue. At the same time, Rockefeller observed the success of Robert Moses, who, through his centrally controlled network of authorities, continued to build public works without ever having to go to the voters. Public authorities offered the governor more than freedom from the requirement of bond referenda, they also allowed the governor the ability to set up agencies of public works construction that bypassed entrenched state bureaucracies of uncreative architects, and the Department of Public Works, with its procrustean relationships with the legislative appropriations process. Rockefeller literally transformed state politics with his own version of the centrally controlled "Moses model" (Perry, 1989). Instead of placing someone like Moses in charge of an aggregate of independent public authorities, the governor, through loyal appointments, put himself in charge. During his 14 years as governor he created 23 new state public authorities, including the Housing Authority (1960), the Job Development Agency (1961), the Metropolitan Transit Authority (1965), the Urban Development Corporation (1968), the Battery Park City Authority (1968), the Mortgage Agency (1970), the Capital District Transportation Authority (1973), and the Medical Care Facilities Finance Agency (1972). Rockefeller used the public authority as the biggest back door in the fiscal history of the state.

Although the governor had employed the Robert Moses approach of centralizing power over public policy in his own office, the authorities he initiated were, in the main, far different from those of Robert Moses. Where Moses's authorities had largely been financially self-sustaining agencies with direct control over the construction, maintenance, operations, and revenue-generating capabilities of the roads and bridges and tunnels, the Rockefeller authorities were just as likely to take on projects traditionally financed and managed by other state government agencies (i.e., construction of the State University and mental hygiene facilities). In fact, "during the later Rockefeller years, public authorities assumed responsibility for projects that were once provided by private enterprise, such as operation of public transportation systems. These projects were *not intended to be self supporting* and thus were to be heavily subsidized by the government and tax payers" (Underwood & Daniels, 1982, emphasis added).

Here the approaches of Rockefeller and Moses were vastly different. Moses combined the public authority with the revenue bond to produce public works with a strong revenue-generating capacity. Rockefeller needed more than the revenue bond to carry out the range of public services he had in mind. He needed a way of involving private-sector lenders in state construction programs that demanded neither voter approval nor an unrealistically high expectation of revenue generation.

The Wall Street bond attorney and future U.S. Attorney General John Mitchell proposed such a way in the indirect form of what came to be called lease-finance arrangements and "moral obligation" bonds. These bonds would be offered through a new type of public authority, which would secure them not only through income from the public works, but also through the moral obligation of the state to make contributions to an authority reserve fund to be used at such times when the authority could not muster sufficent revenues "to cover the payment of principle and interest costs in a given year" (Underwood & Daniels, 1982, p. 172). Critics of this approach have been harsh. It amounted to "unabashed chicanery," observes Daniel Axelrod, "no legislature could bind future legislatures to appropriate funds for a debt reserve" (Axelrod, 1992, p. 5), especially if the debt reserve was made up of tax dollars that would be used to pay off debt that had not been agreed to by voter referendum. At best the bonds represented the state's moral obligation to pay, not its *legal* obligation. Nonetheless, they became an important step at indirectly binding the state to agreements well outside the constitutional requirements of "full faith and credit." Later on John Mitchell is said to have described a jubilant Rockefeller after he apprised the governor of this theory: "Old Nelson espoused my theory like it was the salvation of mankind. He treated the money like manna from heaven" (Glastris, 1985).

Just as important as the moral obligation language on these revenue bonds was another Mitchell design, the lease-financing agreement. Perhaps one of the most innovative backdoor moves in the past three decades, and certainly one that has done as much to increase debt for public works as any other, is the practice of lease financing. Under such agreement a government contracts with or sets up a public authority to build a particular facility. Before construction begins, the authority, which will own the facility, enters into a long-term lease with the leasee-government (the state or city, for example). Using the lease as collateral, the authority floats bonds to finance the construction of the public work. After the facility is completed, the authority collects rent from the government for the use of the facility. The rent is used to retire the principle and interest on the bonds. These rents "result in annual charges against the state's tax revenues or against earmarked revenues used to augment the tax revenues."[16]

It was the political pragmatics of the governor from New York and the wizardry of Wall Street entrepreneurship that combined to write this chapter in the backdoor practices of building the public city. Almost every feature of the single-purpose, revenue-generating, self-sustaining statutory public authority was now overturned. The authority concept, as originally employed and just as originally reemployed by Robert Moses, had been captured by Rockefeller, with the advice of Wall Street, not only for fiscal reasons, but also as an instrument to be used to transform the capital functions of a whole range of public policies. The changes in the scale, breadth of purpose, and financial obligations of state authorities in New York have been profound:

Scale. During the Rockefeller years, the scale of outstanding debt attributable to state authorities increased an astounding 43 times, going from $129 million in 1962 to $5.6 billion in 1972 (Underwood & Daniels, 1982, p. 171). By 1982 the outstanding debt of the 16 largest authorities in New York State was more than $26 billion (New York State Comptroller, 1982), and in 1992 it was in excess of $57 billion

(New York State Comptroller, 1992), or almost 10% of all outstanding state and local public works debt in the United States.

Breadth of Purpose. The mandates given to authorities have also expanded greatly. As late as the 1950s, public authorities were used simply to construct and manage core public works such as highways, bridges, and tunnels. Today they are used to help build every type of capital project, regardless of the public service: from health care and prisons to housing and universities, as well as mass transit, industrial development, and a host of other services.

Financial Obligations. The financial obligations of New York's authorities are a good example of the new range of obligations undertaken by authorities, who, unable to raise enough revenue to service the debt engendered by the broad range of purposes for which they are organized, have become responsible to their more traditional revenue bondholders, but also to bondholders who have lent money under the conditions of moral obligation, contractual obligation, and lease-financing.[17]

Mario Cuomo, long a critic of the undemocratic ways of such institutional politics before he became governor of New York, seemed no less attracted to what has come to be called the "fourth branch of government" once in office. Although he and others had been critical of a government that seemed to be run on bond issues rather than votes, he also conceded that "the authority is a brilliant, brilliant notion. It would be difficult to build bridges or develop ports if voter approval were required for each project." He concluded that "the big advantage [of authorities]: free from the control the people have. The big disadvantage: free from the control the people have" (Bang-Jensen, 1984, p. 4).

As of early 1992 the practice of backdoor finance of public works had reached a new high in New York. Of the $57 billion in total outstanding authority debt, $6.6 billion was in contract obligation debt, $8.3 billion was in lease-purchase debt, and more than $7 billion was in moral obligation debt (see Table 7.2). Although the amount of moral obligation debt allowable to authorities is declining in New York State, the amounts attributable to lease-purchase and contract obligation bonds constitute the fastest growing debt categories. On a per capita basis, 55% of state related-bond debt is now attributable to contract and lease payments (see Table 7.3).

Pennsylvania has aggressively practiced a form of lease-backed financing since 1935, when it began to create local authorities to construct school buildings. These became part of a network of authorities when the state, in the face of strict constitutional restrictions and a tradition of stiff voter approval for bonds, created 2,600 public authorities to provide almost every type of government. Though restrictions in Pennsylvania have been eased somewhat, all educational capital finance is still conducted through a policy of lease-purchase agreements. More recently California and Florida have become innovative practitioners of their own versions of lease-purchase agreements, and Donald Axelrod now estimates that they represent more than "9 percent of all tax-exempt bonds sold. In the next five to ten years this figure should easily triple. At least 41 states have discovered the virtues of lease financing and four—Colorado, Kentucky, South Dakota and Idaho—depend on them almost exclusively to borrow money" (Axelrod, 1992, p. 43).

From Back Door to Trapdoor:
Selling the Public City

At the local level, a disturbing new pattern of backdoor finance is just beginning to emerge in fiscally distressed cities. Instead of agreeing to purchase a newly built facility from a public authority in a lease-backed arrangement, cities like Buffalo and Troy, New York, are actually selling their public works to public authorities for funds to pay off operating budget shortfalls, and then entering into long-term lease agreements with the authorities to pay off the bonds the authorities had to float to purchase the facilities. In Buffalo, the city sold its water system to the managing authority in exchange for $20 million to make up the deficit in its 1992-1993 annual budget. In Troy, New York, the city sold its city hall and seven other city-owned capital facilities, including a city park, to the local development corporation (LDC) for $35 million. The LDC raised the funds for the purchase through the sale of revenue bonds, and now the City rents city hall and the city park from the LDC. The City used $10.5 million dollars of the funds to pay off a shortfall in its annual working capital budget.

Critics, including the Comptroller of the State of New York, have argued that deals like the one in Troy are "an elaborate and purposeful circumvention of State Law and policy" (Brown, 1993, p. 1). The architects of this strategy, including the city manager and several city council members, have "consistently defended the deal, saying it was an alternative to raising taxes and cutting services when state aid is dwindling in tight fiscal times" (Brown, 1993, p. 1).

TABLE 7.2 Net Tax-Supported Debt: Defined by Traditional Category (in Millions of Dollars, Outstanding as of March 31, 1992)

Short-Term Operating Debt[1]		$3,900
General Obligation Debt[2]		5,081
Contractual-Obligation Debt[3]		
Dormitory Authority[4]	$18	
Housing Finance Agency[5]	337	
Local Government Assistance Corporation[6]	2,701	
Medical Care Facilities Finance Agency[7]	590	
Metropolitan Transportation Authority[8]	2,004	
Thruway Authority[9]	220	
Triborough Bridge & Tunnel Authority[10]	519	
Urban Development Corporation[11]	225	
Total		6,604
Lease-Purchase Debt[3]		
Certificates of Participation[12]	623	
County of Albany-Empire State Plaza[13]	334	
Dormitory Authority[14]	5,148	
Environmental Facilities Corporation[15]	101	
Other Office Buildings[16]	9	
Thruway Authority[17]	306	
Urban Development Corporation[18]	1,802	
Total		8,325
Moral Obligation Debt[3, 19]		7,008
Other Debt[20]		4
Total Net Tax-Supported Debt		$24,630

SOURCE: NYS Comptroller (1992).

1. Repaid on March 31, 1992. Excludes $531 million of outstanding tax and revenue anticipation notes sold on March 25, 1992.

2. Consists of $4,743 million of bonds and $338 million of commercial paper and bond anticipation notes. Excludes $177 million of $405 million of State General Obligation Refunding Bonds dated August 1, 1986 to advance refund an equal amount of general obligation bonds that are included in this table. Debt service on the refunding bonds will be paid from an escrow fund established with the proceeds of the refunding bonds until the refunded bonds are redeemed from the balance of the proceeds of the escrow fund.

3. Payments made by the State used to pay debt service on contractual obligations, lease-purchase obligations, and moral obligations are subject to annual apropriations by the Legislature and the availability of moneys to the State for the purposes of making such payments.

4. Issued to finance a library for the blind and handicapped.

5. Issued to finance various housing programs.

6. Issued to finance certain assistance payments to local governments; secured by a portion of certain sales and use taxes.

7. Issued to finance certain hospital projects; secured by service contracts providing for payments by the State if necessary, subject to annual appropriations for up to 30 years.

8. Issued to finance transit improvements; secured by service contracts providing for payments by the State over a 35-year period.

TABLE 7.2 (Continued)

9. Issued to finance the Consolidated Highway Assistance Program.
10. Issued to finance construction of the New York City Convention Center.
11. Issued to finance construction of high technology centers at certain universities, Ten Eyck Plaza in Albany and the Onondage County Convention Center.
12. Issued to finance the acquisition of certain equipment and facilities.
13. Issued to finance the State's office complex in Albany. Of this amount, $211 million relates to debt issued by Albany County for construction of the complex and $123 million relates to lease rental bonds issued by the Urban Development Corporation to restructure debt service payments totalling $126.7 million in fiscal years 1989,1990,1991 on the Albany County bonds.
14. Issued for five purposes: (i) $2,162 million to finance capital improvements at the City University of New York; (ii) $2,791; (iii) $27 million for athletic facilities at SUNY; (iv) $24 million for the State Education Department; and (v) $144 million for the Department of Health. Of the amount for capital improvements at SUNY, $1,809 million relates to obligations for Senior Colleges on which the State contributes funds that pay 100% of debt service and the remainder ($353 million, equal to 50% of the total of $706 million) relates to obligations for community colleges on which the State contributes funds that pay 50% and New York City pays 50% of debt service.
15. Issued to finance Riverbank Park and the State's share of the contribution to the Corporation's revolving loan fund.
16. Issued to finance construction of State office buildings.
17. Issued to finance highway reconditioning, preservation, construction, and reconstruction.
18. Issued to finance correctional facilities and state capital construction costs.
19. Issued by the Urban Development Corporation (UDC) to finance housing projects: There are no assurances that these programs can remain self-sustaining in view of their past difficulties. For that reason, the Comptroller's Office has made a judgment that UDC debt issued (and outstanding) for housing programs should be included in New York's net tax-supported debt position.
20. Issued by Erie County to finance its convention center. The amount of debt shown in the table is the portion for which the State contributes debt service (50% of $7.5 million).

It is somewhat ironic to contemplate what is going on here. The whole notion of debt finance could be about to take another curious turn in these perilous financial times: where the local working capital debt will be paid off (at least once) through the sale and lease of the very infrastructure the locality went into debt once before to build. When faced with another working capital shortfall, what else will Troy sell? It can only sell city hall once. To call this form of fiscal circumvention the politics of the back door might be a misnomer; more likely, it might be the first stages of a politics of the trapdoor—where the physical infrastructure policy of the city further erodes the political-economic stability of an already fragile urban environment.

■ Conclusion

The politics of building the public city in Troy, New York, is a far cry from the visions of Andrew Gallatin or the power brokering of Robert Moses. Building public infrastructure has been an uneven prac-

TABLE 7.3 State-Related Debt per Capita

Fiscal Year	Total State Population[1] (Millions)	General Obligation Debt[2]	State-Guaranteed Authority Debt	Lease-Purchase and Contractual-Obligation Debt[3]	Moral Obligation Debt	Total Debt
1974-75	18.1	$194	$31	$199	$283	$707
1979-80	17.6	206	30	222	656[5,6]	1,114
1984-85	17.7	215	35	318	726[5,7,8]	1,294
1988-89	17.9	244	31	466	647[5,7,9]	1,388
1989-90	18.0	278	31	519	613[5,7,10]	1,441
1990-91	18.0	318	30	735[4]	550[5,11]	1,599
1991-92	18.1	281	28	960[4]	523[5,12]	1,735

SOURCE: NYS Comptroller (1992).

1. For calendar year ending in State fiscal year.
2. Excludes tax and revenue anticipation notes.
3. In years prior to 1988-89 this category includes two Housing Finance Agency programs (State University of New York and Mental Hygeine), which also had the moral obligation provision. In the 1988-89 fiscal year, and thereafter, included in this category is one HFA program (State University of New York), which also has the moral obligation provision. Included in the 1984-85 and subsequent fiscal years is indebtedness of the Dormitory Authority issued to finance educational construction for the City University of New York (Senior Colleges) for which the State contributes, subject to annual appropriation, to rental payments equal to 100% of the debt service. Included in the 1990-91 and 1991-92 fiscal years are certificates of participation that represent proportionate interests in lease payments to be made by the State.
4. Includes $708 million and $706 million, respectively, of obligations issued by the Dormitory Authority for City University of New York Community Colleges for which the State pays 50% of the debt service.
5. Includes $6.182 billion, $8.110 billion, $7.537 billion, $7.122 billion, $6.705 billion and $5.338 billion, respectively, of outstanding MAC debt.
6. Moral obligation debt excluding MAC was $5.390 billion, or $305 per capita.
7. Includes $316 million, $270 million, and $263 million, respectively of outstanding debt of the Dormitory Authority for the City University of New York for which the State contributes, subject to annual appropriations, to rental payments in amounts equal to 50% of the debt service.
8. Moral obligation debt excluding MAC was $4.769 billion, or $269 per capita.
9. Moral obligation debt excluding MAC was $4.039 billion, or $226 per capita.
10. Moral obligation debt excluding MAC was $3.922 billion, or $218 per capita.
11. Moral obligation debt excluding MAC was $3.200 billion, or $178 per capita.
12. Moral obligation debt excluding MAC was $3.122 billion, or $172 per capita.

tice, including everything from entrepreneurial vision and social reform to constitutional intrigue and fiscal desperation, all the while characterized by the central tension between the demand for the public works and the ability to pay for it.

The purpose of this chapter has been to show how building the public city is very much an issue of finance; more precisely, of debt. The politics of public infrastructure is very much about *who* will pay for these complex technological networks, buildings, and other facilities. Or again, more exactly, who will bear the debt. The ways in which the federal, state, and local governments have historically set out to solve these issues form essential parts of the day-to-day practice of what passes for public infrastructure policy in the United States.

Today much has been made of the deplorable state of much of the nation's public infrastructure and the need for new national policies and fiscal commitments to solving some of the complications evidenced by such problems (Choate & Walter, 1983; NCPWI, 1988). However, if history is in any way instructive here, it is that the real problems with public infrastructure delivery are most keenly felt at the state and local level, and it is there, as well, where the most flexible conditions of reform will probably be best realized. The lion's share of all public infrastructure efforts is at the state and local level, where localities are responsible for the management and maintenance of more than 70% of the nation's public works and services (OTA, 1990, p. 1). At the same time there has been a significant withdrawal of federal aid to states and localities for public infrastructure, and there is limited evidence to suggest any major reversal of this trend. So the bond-indebtedness policies of states and especially localities, who now finance two thirds of the capital needed for infrastructure in this country, will in all likelihood continue to mushroom. Add to this the combined conditions of state limits on debt, the increased number of federal and state unfunded mandates, grant requirements (such as the focus on new construction in earmarked grants rather than on maintenance), and regulatory changes (all of which were not discussed directly in this chapter), and the result is a public infrastructure process that is becoming more constrained even as it is becoming more overburdened.

For example, at this point in time New York State is one of the most overburdened and highly taxed states in the United States, using a full range of debt-creating and tax-evading devices that have still have not been enough to stop a slippage in its debt rating that began in 1984. Although the increasing use of new refinements of backdoor finance for public works has allowed the state to continue to build public works, it is beginnng to have deleterious effects: The state rests at or near the bottom of most credit rankings[18] and its debt position has, in Wall Street terms, worsened steadily. The state's cities, as discussed here, are taking backdoor financing to new depths of fiscal desperation.

All this suggests that the present practices of backdoor financing of public works have the potential to wear out. The clearest and most basic two avenues of major change are to switch the responsibility for some public services to another level of government, or to change the conditions under which we pay for public infrastructure. The former choice means rethinking the federal role in providing public infrastructure— something that has only been done in the most sporadic of ways in the past 200 years. The latter choice means rethinking how we directly tax ourselves for public works. At all levels of government, politicians and voters might find that they will need to reconsider the way they directly provide for the large-scale "technological sinews" that make up the public city. Otherwise, more and more cities will find themselves practicing the politics of Troy: no longer building the public city, but selling it, just to make ends meet.

NOTES

1. Lindsay v. Comm'rs, 2 Bay 38, 46, 53 (S.C., 1796), as discussed in Schultz (1989, p. 38).

2. See Schultz (1989, pp. 35-111) for a more complete discussion of the regulation of public infrastructure in the urban environment.

3. The federal government was not totally inactive. On the contrary, historians like Bruce Seeley (1993) and George Rogers Taylor (1951) both suggest that the federal government, despite its constitutional reticence, provided the occasional grant for the construction of a turnpike, in states like Ohio or Indiana, or the purchase of canal stock. The government also made use of eminent domain and gave millions of acres to canal and railroad projects.

The federal government also make its presence known through the U.S. Army Corps of Engineers, which was responsible for surveying various transportation sites and also maintaining and clearing rivers and harbors.

4. In a footnote in his classic study on public debt, Henry Adams reports that "previous to 1880 Congress had granted 215,000,000 acres of land to railroads and canals, of which titles were secured to 150,000,000 acres" (Adams, 1892, p. 356).

5. See Henry Adams (1892, p. 324) for a fuller discussion and quotation.

6. Although, throughout the century, states did receive funds from the federal government for occasional projects. For the most part the federal government confined its role to the taking of land and to the provision to the states, for a while during the 1830s, of

some money, in the form of a revolving fund established from surplus revenues in the national treasury and a percentage of the earnings on federal land sales. Henry Adams (1892, p. 327) argues that rather than stimulate responsible state activity in internal improvements, this largesse from the federal government had the opposite effect generating a "carelessness with which local obligations were incurred."

7. New York State Controller Flagg in 1842, as quoted in Quirk and Wein (1971, p. 526).

8. From the Constitution Convention Debates of 1846, as abstracted by Quirk and Wein (1971).

9. See letter of Franklin D. Roosevelt to the governors of all the states, as quoted in Smith (1974) and Shestack (1957, pp. 553-569).

10. See the Robert Moses Archives at SUNY at Buffalo.

11. This is clear in his correspondence with governor-elect Franklin Roosevelt and his drafts of the legislation to create the Triborough Bridge Authority in 1928 (Moses, undated, 1933a, 1933b).

12. See *The Revised Record of the New York State Constitutional Convention of 1938* (NYS CCC, 1938b). Comments of Alfred Low Moffatt, 2259.

13. See *The Revised Record of the New York State Constitutional Convention of 1938* (NYS CCC, 1938b). Comments of Robert Moses, 2264 and 2266.

14. See *The Revised Record of the New York State Constitutional Convention of 1938* (NYS CCC, 1938b). Comment of Governor Alfred Smith, 2268.

15. See James Liegland (this volume) as well as Advisory Commission on Intergovernmental Relations (ACIR, 1986).

16. This observation is found in the State Comptroller's *Annual Report 1973*, p. 19, as quoted in Underwood and Daniels (1982, p. 170).

17. For a more complete discussion of each of these categories of debt and their various distinctions, see the New York State Comptroller (1992). For example, not every authority can issue moral obligation bonds in New York—only the Housing Finance Agency, the Dormitory Authority, the Medical Care Facilities Agency, the Urban Development Corporation, the Project Finance Agency, the Battery Park City Authority, the United Nations Development Corporation, and the Municipal Assistance Corporation of the City of New York are able the raise money via "MOs" and then only with the direct permission of the legislature. Other authorities can raise money via "contract obligation" agreements with the state, whereby the state makes periodic payments to the authority for the service of a specified amortized period of debt to bondholders who have lent money for a particular service. These differ from lease-financing arrangements, which are centered on a lease arrangement of a particular service by the state or a municipality. In the case of New York State, the primary public works it leases from authorities are medical health facilities, university dormitories and buildings, correctional facilities, highways, and state and municipal office buildings.

18. At the current time, the general obligation debt of New York State is rated "A" by Moody's, ranking it 49 out of 50 states. The bonds are rated "A-" by Standard and Poors, ranking it last among the 50 states. The TRANS (Tax and Revenue Anticipation Notes) are also rated near the bottom of the 50 states by Fitch, Moody, and Standard and Poors. These rankings have not changed much since 1984. The calculation of indebtedness used most frequently by the bond houses is the one that takes into account all issues supported by the state's tax revenues—both those constitutionally recognized under conditions of "full faith and credit," and those indirectly supported through arrangements of moral

obligation and contract, or the like. The Office of the State Comptroller of New York (1992) reports that the debt position of New York State continues to worsen as more and more debt, for which the state holds some form of tax-related obligation, continues to accumulate.

REFERENCES

Adams, H. (1892). *Public debts: An essay on the science of finance.* New York: D. Appleton.

Advisory Commission on Intergovernmental Relations (ACIR). (1986). *Significant features of fiscal federalism* (1985-86 Ed.). Washington, DC: Government Printing Office.

Anderson, L. (1988). Fire and disease: The development of supply systems in New England, 1970-1990. In J. A. Tarr & G. Dupuy (Eds.), *Technology and the rise of the networked city in Europe and America* (pp. 137-156). Philadelphia: Temple University Press.

Axelrod, D. (1992). *Shadow governments: The hidden world of public authorities—And how they control over $1 trillion of your money.* New York: John Wiley.

Bang-Jensen, L. (1984). *New York State's other government: The long shadow of public authorities* (The Rockefeller institute special report series). Albany, NY: The Rockefeller Institute.

Brown, N. P. (1993, November 18). Comptroller blasts Troy: Leaseback "circumvents state law." *The Record,* p. 1.

Callender, G. S. (1903). The transportation and banking enterprises of the states in relation to the growth of corporations. *Quarterly Journal of Economics, 18,* 151-152.

Caro, R. A. (1974). *The power broker: Robert Moses and the fall of New York.* New York: Knopf.

Chevalier, M. (1839). *Society, manners and politics in the United States, being a series of letters on North America.* Boston: Weeks, Jordon and Company.

Choate, P., & Walter, S. (1983). *America in ruins: The decaying infrastructure.* Durham, NC: Duke Press Paperbacks.

Council of State Governments (CSG). (1953). *Public authorities in the states* (mimeo). Lexington, KY: Author.

Council of State Governments (CSG). (1970). *State public authorities.* Lexington, KY: Author.

Davis, H. A. (1935, June). Borrowing machines. *National Municipal Review,* pp. 328-334.

Glaab, C. N., & Brown, A. T. (1967). *A history of urban America.* New York: MacMillan.

Glastris, P. (1985, February). The government debt racket. *Washington Monthly,* pp. 12-24.

Goodrich, C. (1960). *Government promotion of canals and railroads, 1800-1890.* New York: Columbia University Press.

Isard, W. (1942). A neglected cycle: The transport building cycle. *Review of Economic Statistics, 24,* 149-158.

Lincoln, C. Z. (1900). *The constitutional history of New York* (Vol. 2). Rochester: Lawyers Cooperative.

Lindsay v. Comm'rs, 2 Bay 38, 46, 53 (S.C., 1796).

Lines, J., Parker, E., & Perry, D. (1986). Building the twentieth century public works machine. In M. Schoolman & A. Magid (Eds.), *Reindustrializing New York State: Strategies, implications, challenges* (pp. 231-256). Albany: SUNY Press.

Moses, R. (undated). [Moses to Roosevelt]. Robert Moses Research Archives, SUNY/ Buffalo.

Moses, R. (1933a). [Moses, Coombs, Battle correspondence]. Robert Moses Research Archives, SUNY/Buffalo.

Moses, R. (1933b). [Moses to Hoey]. Robert Moses Research Archives, SUNY/Buffalo.

Moses, R. (1941). [Opening remarks on the occasion of the fifth anniversary of the opening of the Triborough Bridge]. Robert Moses Research Archives, SUNY/Buffalo.

National Council on Public Works Improvement (NCPWI). (1988). *Fragile foundations: A report on America's public works. Final report to the president and Congress.* Washington DC: Government Printing Office.

New York State Comptroller. (1982). *Annual Report 1982.* Albany, NY: Author.

New York State Comptroller. (1992). *$2,300,000,000 state of New York: 1992 tax and revenue anticipation notes* (Official statement). Albany, NY: Author.

New York State Constitution. (1938). Article VII, Section 12.

New York State Constitutional Convention Committee (NYS CCC). (1938a). *Problems relating to taxation and finance.* Albany, NY: J. B. Lyon.

New York State Constitutional Convention Committee (NYS CCC). (1938b). *Revised record of the New York State constitutional convention of 1938.* Albany, NY: Author.

New York Statutes. (1933). Chapter 145.

New York Statutes. (1937). Chapter 3.

Perry, D. C. (1989). The Moses model of governance. In J. Krieg (Ed.), *Robert Moses: Single minded genius.* Interlaken, NY: Heart of Lakes Press.

Perry, D. C. (1990a). *Building the city with authority* (Albert A. Levin working papers series). Cleveland, OH: Cleveland State University.

Perry, D. C. (1990b). *Robert Moses, the public authority and the bridge that became a tunnel.* Paper presented at the *Robert Moses' New York* conference. Columbia University, New York.

Quirk, W. J., & Wein, L. E. (1971). A short constitutional history of entities commonly known as authorities. *Cornell Law Review, 56*(4).

Samuelson, R. (1990, January 3). A frivolous debate? *Washington Post*, p. A15.

Schiesl, M. J. (1977). *The politics of efficiency: Municipal administration and reform in America, 1800-1920.* Berkeley: University of California Press.

Schultz, S. K. (1989). *Constructing urban culture: American cities and city planning, 1800-1920.* Philadelphia: Temple University Press.

Seeley, B. (1993, Winter). The saga of American infrastructure. *Wilson Quarterly*, 21-39.

Segal, H. H. (1961). Cycles of canal construction. In C. Goodrich (Ed.), *Canals and American economic development* (pp. 169-215). New York: Columbia University Press.

Shestack, J. (1957). The public authority. *University of Pennsylvania Law Review, 105*, 553-569.

Smith, R. (1974). *Ad-hoc governments: Special purpose transportation authorities in the United States and Britain.* Beverly Hills, CA; Sage.

Smith, R. C. (1964). *Public authorities, special districts and local government.* Washington DC: National Association of Counties Research Foundation.

Spargo, G. (1983). Interview for the Robert Moses Archives, SUNY/Buffalo.

Tarr, J. A., & Dupuy, G. (Eds.). (1988). *Technology and the rise of the networked city in Europe and America.* Philadelphia: Temple University Press.

Taylor, G. R. (1951). *The transportation revolution: 1815-1860.* New York: Rinehart.

Tobin, A. J. (1953, March 26). *Authorities as a governmental technique.* Paper presented at Rutgers University.

Underwood, J. E., & Daniels, W. J. (1982). *Governor Rockefeller in New York: The apex of pragmatic liberalism in the United States.* Westport, CT: Greenwood.

U.S. Congress, Office of Technology Assessment (OTA). (1990). *Rebuilding the foundations: State and local public works financing and management.* Washington, DC: Government Printing Office.

Walsh, A. H. (1978). *The public's business: The politics and practices of government corporations.* Cambridge: MIT Press.

West River Bridge Co. v. Dix, Vt, 47 US 507, 6 HOW 507, 12 Led 829 (1847, US Supreme Court).

Name Index

Subject Index

About the Contributors

JAMESON W. DOIG is Professor of Politics and Public Affairs at the Woodrow Wilson School and the Department of Politics at Princeton University. He is coauthor of *Leadership and Innovation: Entrepreneurs in Government* (1990) and has also written widely in the fields of urban development, government organization, and criminal justice.

CLAIRE L. FELBINGER is Associate Professor of Public Administration and Urban Studies at the Levin College of Urban Affairs, Cleveland State University. She is also Director of the College's Master's of Public Administration program, which includes the nation's largest graduate program in public works management. Her research has been widely published in books and articles on public works management, economic development, urban service delivery, intergovernmental relations, and evaluation research methodology.

JAMES LEIGLAND has worked since the early 1980s as a member of the internationl research and consulting staff of the Institute of Public Administration, New York City. He has written extensively on public finance topics, including municipal bonds and public authorities, and is coeditor of the *Handbook of Municipal Bonds and Public Finance*, published by the New York Institute of Finance. He is currently completing a technical assistance assignment as a senior resident advisor to the Indonesian National Development Planning Agency.

ROBERT MIER is Professor of Urban Planning and Public Administration at the University of Illinois at Chicago. From 1983 to 1989, he was the Commissioner of Economic Development and the Deputy Mayor for Development, during the tenures of Chicago Mayors Harold Washington and Eugene Sawyer. He was the architect of Chicago's acclaimed *1984 Development Plan*. Mier is author of the recently published *Social Justice and Local Development Policy* and the coeditor of *Economic Development in the United States*.

249

DAVID C. PERRY is Professor of Planning at the School of Planning and Architecture at SUNY Buffalo. His research and teaching interests include urban political economy, planning theory and policies of infrastructure, economic development, and community change. His books include *Rise of the Sunbelt Cities,* the forthcoming *Spatial Practices: Critical Explorations in Social/Spatial Theory,* and *The Cleveland Metropolitan Reader.*

HEYWOOD T. SANDERS is Professor of Urban Administration at Trinity University in San Antonio, Texas. His publications include *The Politics of Urban Development* (with Clarence Stone) and *Urban Texas: Politics and Development* (with Char Miller), as well as articles in *Urban Affairs Quarterly*, the *Journal of Urban Affairs*, and *The Public Interest*. He is currently completing a book on the politics of urban infrastructure and capital investment.